"This timely and outstanding book, written by a respected commentator and acknowledged expert on casino tourism, is essential reading both for specialists in the field and those interested in the broader touristic significance of a growing and economically important business. The specific focus on east and southeast Asia offers uniquely detailed insights into an area of the world where tourism is evolving rapidly. The author provides a wealth of data on, and analysis of, the role being played by casino tourism in this process of evolution. Highly recommended."

– **Dr Roy C. Wood**, Visiting Professor, Hertfordshire Business School, University of Hertfordshire, UK.

"I would highly recommend this book to anyone interested in delving into the captivating world of casino tourism, particularly in East and Southeast Asia. As someone who is always been fascinated by this subject, I was thrilled to discover that the author, a renowned figure in the field, has poured extensive knowledge and insights into its pages. The book not only presents ten thought-provoking research questions that guide readers through the evolution of casino tourism in the region, but it also introduces an ingenious analytical framework that offers a fresh perspective on its development. For me, this book has been an enlightening exploration, shedding new light on the world of casino tourism in East and Southeast Asia."

– **Dr Amy Siu Ian So**, Associate Professor, Department of Integrated Resort and Tourism Management, University of Macau, China

"Casino tourism boomed in East and Southeast Asia after Macau opened the sector to competition in 2004. Many other jurisdictions soon followed, including Singapore, in the context of a turn to an integrated resort model, with each jurisdictions adopting its own variation. This book discusses the key policy choices and compares various regulatory models, institutional settings, and their economic and social implications. It is an invaluable resource to understand modern regulatory approaches to the casino tourism industry in Asia."

– **Dr Jorge Godinho**, Associate Professor, Instituto Superior Manuel Teixeira Gomes (ISMAT), Portimão, Algarve, Portugal

Developing the Frontiers of Casino Tourism

This book explicates how the frontiers and path of development of casino tourism in various destinations in East and Southeast Asia are shaped by a set of social, cultural, political and economic forces and their interplays in a dynamic environment, and the prospects of this industry in different destinations in the 2020s and onward.

Casino tourism has been a rapidly growing industry in East and Southeast Asia in the 2000s and 2010s. By scrutinizing the respective evidence from the better-developed, emerging and potential casino tourism destinations in East and Southeast Asia (which include Macao, Singapore, Malaysia, South Korea, the Philippines, Vietnam, Cambodia, Myanmar, Japan, Thailand and Taiwan), the book analyses specific contextual dimensions, including social, cultural, political and economic standpoints; the pragmatic features associated with the progress of this industry; and performances of casino tourism at the various East-Southeast Asian casino tourism destinations. Additionally, it also explores the impacts of the "China factor" and the resultant changes to the development frontiers of casino tourism. Thus, readers who would like to gain realistic insights into the underlying forces and dynamics of the practices and prospects of this industry will find this book informative and compelling.

This book provides a comprehensive and succinct reference to students and industry practitioners in the areas of tourism, casino gaming and integrated resorts who wish to gain further insights into the general and essences, business routines and prospects of casino tourism development in East-Southeast Asia. Academics may also find this book a useful source for further exploration.

Ricardo C.S. Siu is Associate Professor of Business Economics in the Faculty of Business Administration at the University of Macau. His research work focuses on institutional economics, economics of casino gaming/tourism and the Chinese economy. Ricardo is an internationally known scholar in the stream of casino gaming, specializing in the areas of public policy and evolution of casino gaming in Macao and other Asian jurisdictions. He is a frequently invited speaker to various executive development programs in gaming management. Ricardo has widely published in academic journals, books and professional magazines in the areas of casino gaming and tourism. He is the author of the book *Economic Principles for the Hospitality Industry* (2020).

Routledge Insights in Tourism Series

Series Editor: Anukrati Sharma

Head & Associate Professor of the Department of Commerce and Management at the University of Kota, India

This series provides a forum for cutting edge insights into the latest developments in tourism research. It offers high quality monographs and edited collections that develop tourism analysis at both theoretical and empirical levels.

Developing Industrial and Mining Heritage Sites
Lavrion Technology and Cultural Park, Greece
Taşkın Deniz Yıldız

Innovation Strategies and Organizational Culture in Tourism
Concepts and Case Studies on Knowledge Sharing
Edited by Marco Valeri

Tourism and Poverty Alleviation in Nature Conservation Areas
A Comparative Study Between Japan and Vietnam
Nguyen Van Hoang

Sport Tourism, Events and Sustainable Development Goals
An Emerging Foundation
Edited by Anukrati Sharma, Miha Lesjak and Dusan Borovcanin

Tourism, Philanthropy and School Tours in Zimbabwe
Problematising "Win-Win" Discourses
Kathleen Smithers

Metaverse and Tourism
Rethinking Implications on Virtual Reality
Edited by Marco Valeri and Ahmad Albattat

Developing the Frontiers of Casino Tourism
Progress and Prospects in East and Southeast Asia
Ricardo C.S. Siu

For more information about this series, please visit: www.routledge.com/Routledge-Insights-in-Tourism-Series/book-series/RITS

Developing the Frontiers of Casino Tourism

Progress and Prospects in East and Southeast Asia

Ricardo C.S. Siu

Routledge
Taylor & Francis Group
LONDON AND NEW YORK

First published 2025
by Routledge
4 Park Square, Milton Park, Abingdon, Oxon OX14 4RN

and by Routledge
605 Third Avenue, New York, NY 10158

Routledge is an imprint of the Taylor & Francis Group, an informa business

British Library Cataloguing-in-Publication Data
A catalogue record for this book is available from the British Library

ISBN: 978-1-032-24718-2 (hbk)
ISBN: 978-1-032-24719-9 (pbk)
ISBN: 978-1-003-27988-4 (ebk)

DOI: 10.4324/9781003279884

Typeset in Times New Roman
by Apex CoVantage, LLC

Contents

Dynamics and Concerns

Cautious Planning and Different Practices

Active Participants and Imitators

Figures

Tables

Acknowledgment

I would like to express my heartfelt thanks to my employer, the University of Macau, for providing me with opportunities to focus on my research work in casino gaming and casino tourism which have resulted in the success of related works and the completion of this book. In addition, the patience and efforts of Faye Leerink, Prachi Priyanka and Andy Humphries at Taylor & Francis/Routledge in assisting me to meet the requirements throughout the publication process of this book are greatly appreciated.

Also, I must thank an anonymous reviewer for the very objective and constructive comments on my proposal of this book, which largely inspired me to rethink the organization and flow of materials. Lastly, I would like to express my sincere gratitude to my wife, Lina Chan, and children, Charmaine Siu and Chapmann Siu, who gave their wholehearted support to me during the writing of this book.

Ricardo C.S. Siu
April 20, 2024

1 Introduction

The development of casino tourism in East and Southeast Asia (ESEA) in the first two decades of the twenty-first century has captured the attention of not only the related business firms and investors around the world but also governments, local communities and the academia (see, e.g., Ashton and Korpi (2008); Dolezal et al. (2020); Gunesch (2017); Lee (2017); McCartney (2016); Siu (2013); and Zhang (2017)). The industry has been conspicuous in its performance, as evidenced by the growth of its gross gaming revenue (GGR) and non-gaming tourism sector within all-encompassing destinations like mega casino resorts or integrated resorts (IRs), as well as the business opportunities that are extended to other local tourism sectors. Aside from the increased income and spending power of ESEA visitors which have subsequently increased the demand for casino tourism and advanced the casino tourism industry, it is important to note that the related policies approved by the various governments at the outset promote capital investments in both gaming and non-gaming facilities. The capital investments contribute to increases in hospitality capacity, attraction of the destination and quality of the service in the destination itself. During the process, the pioneering and innovative business strategies of industry practitioners to stimulate demand further expand the market.

Nevertheless, in view of the possible downsides associated with the practice of casino gaming, especially both reported and presumed social issues related to gambling addiction, crime, corruption and so on and so forth, new approvals for this business sector and its continuous expansion are often met with social resistance. Thus, effective public policies and business decisions that are attuned to the specific societal contextual settings and interests of the general public are necessary or even a prerequisite to ensure the sustained growth of casino tourism. Besides, considering that casino tourism is innately a socially controversial industry, whether and when this market matures or even becomes saturated in any single destination after a period of rapid growth would definitely be a crucial topic to industry stakeholders.

To address these issues, it is important to understand the specific factors and dynamics in the ESEA that contribute to the performance of the different destinations and their changes over time, differences in performance among the destinations, and ongoing concerns and prospects of this industry. Doing so allows industry practitioners to gain insights that push the frontiers of casino tourism to

DOI: 10.4324/9781003279884-1

advance and promote their businesses, and formulate effective business strategies accordingly. Besides, a good understanding of these factors and dynamics is necessary for governments to make effective and proactive changes to their industrial policies that sustain the growth and development of this industry, hence minimizing its potential adverse impacts while optimizing its contributions to the economy and society. In addition, appropriate knowledge of the particular contextual settings of ESEA and their impacts on the development of casino tourism in this region may also allow academics to pursue practical research work on related topics.

The first section of this chapter describes the geographical area and current composition of casino tourism in the ESEA as the introduction component of this book. Then, the evolving performance of this industry in various destinations based on business turnovers in casino gaming and international tourism expenditures is discussed in the section that follows. The identified facts and evidence are then used to structure the major inquiries that would further the frontiers of casino tourism in the third section. To examine the factors that push the frontiers, an analytical framework is proposed and elaborated in the fourth section. Then, a brief outline of this book is presented in the last section.

Takeaways from this chapter:

1. The composition of casino tourism in the ESEA is unique to its geographical and contextual settings.
2. Casino tourism in the ESEA continues to evolve and grow.
3. The modern development of casino tourism in the ESEA has its own dynamics which warrants an in-depth investigation.
4. The growth in casino gaming and non-gaming tourism expenditures is highly correlated in the different ESEA destinations.
5. As a socially controversial industry, the laws and legal system at the outset largely frame the structure and business organization for the development of the casino gaming industry and its interplays with tourism.
6. The business performance of casino tourism is not simply driven by market forces but also involves the trade-off among the social, cultural, political and economic forces in a society.

1.1 Composition of casino tourism in East and Southeast Asia

The United Nations defines East Asia (EA) has five countries and one *de facto* country, which are China (including Mainland China and its two special administrative regions (SARs) of Hong Kong and Macao), Japan, Mongolia, North Korea, South Korea and Taiwan (as shown in the dark-gray area in Figure 1.1). On the other hand, Southeast Asia (SEA) includes 12 countries: Andaman & Nicobar Islands, Brunei, Cambodia, East Timor, Indonesia, Laos, Malaysia, Myanmar, the Philippines, Singapore, Thailand and Vietnam (as shown in the light-gray area in Figure 1.1).

In terms of geographical area, it is a well-known fact that China dominates the ESEA (also evident from Figure 1.1), accounts for over 50% of the total

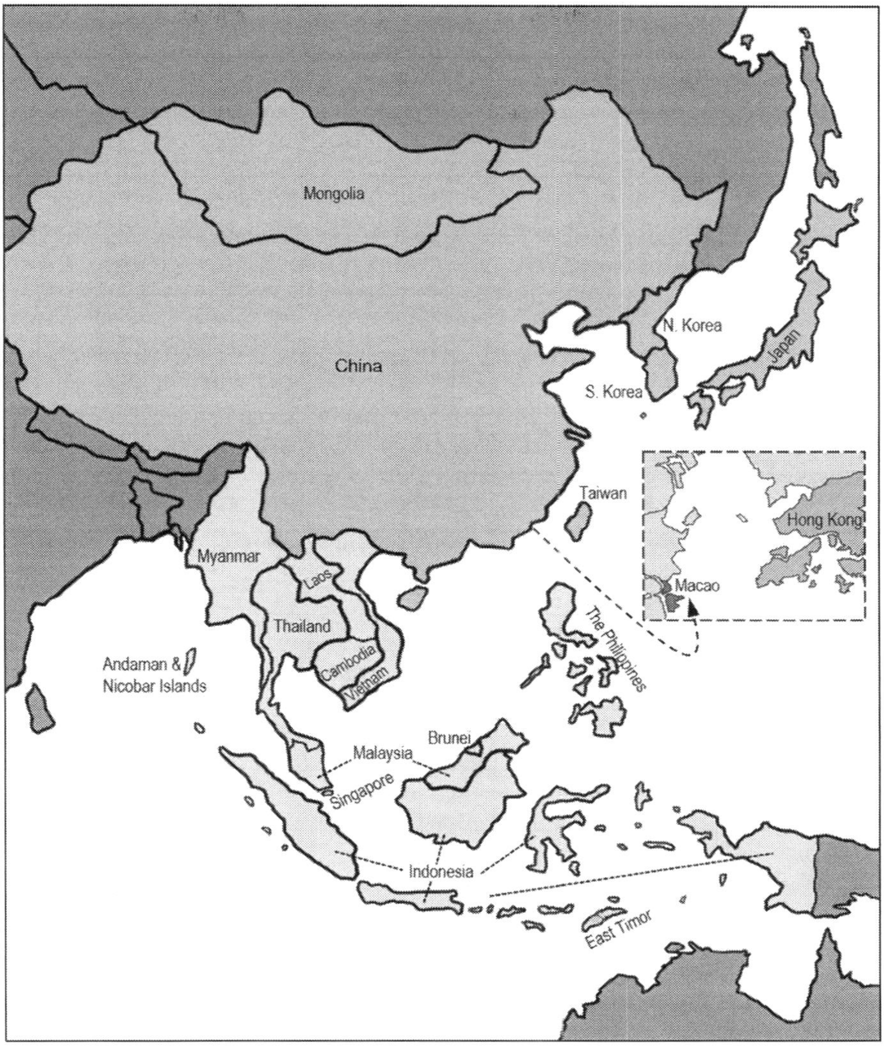

Figure 1.1 Geographical composition of East and Southeast Asia

Source: Tochi Leung

geographical area of the ESEA and 80% of the EA and is twice the size of the SEA. Besides, with a population of 1.4 billion in 2023, China represents around 60% the total population of the ESEA and 87.5% of that of the EA, or it has double the population of the SEA. As the second largest economy behind the USA in terms of gross domestic product (GDP) since 2010, the success of the Chinese economy has undoubtedly increased the real purchasing power and propensity of its people to spend on luxury goods and services, which include casino tourism. Thus, Chinese outbound tourists considerably contribute to business sectors like retail and entertainment in tourism destinations in the ESEA.

Besides, gambling has always been illegal in Mainland China, while casino tourism was given approval to continue after the handover of Macao from the Portuguese to the Chinese after the establishment of the Macao SAR at the end of 1999 under the "one-country, two-system" policy. The dramatic growth and development of the casino industry in Macao from 2000 to 2020 undoubtedly caught the attention of other destinations in the ESEA on the remarkable potential of this business sector for Chinese patrons. Besides, the outstanding economic growth in the ESEA also heightened industry and government expectations towards the demand for casino tourism in this part of the world. Thus, it is observed that existing casino tourism destinations like Cambodia, the Philippines and Vietnam are actively moving to expand the scale of their industries (see, e.g., Barta (2012); Foster and Dan (2022); and Zhu (2022)). Besides, casino gaming has also been legalized in other countries like Singapore in 2005 (China Daily 2005) and most recently in Japan in 2018 (O'Connor 2021) to further boost the development or redevelopment of their tourism industries and generate positive impacts to the other parts of their economies. The legal status of casino tourism in the different countries/regions in the ESEA at the end of 2023 is summarized in Table 1.1.

Based on the information in Table 1.1, it is reasonable to say that casino tourism is still in its infancy in different countries for the coming decade. For example, the earliest that Japan is expected to open its first IR is 2030 (Chan 2023) and the Japanese government may also grant two more casino gaming licenses in 2025 or later. In addition, while investors are still waiting for possible opportunities in Taiwan, other governments are now designating casino gaming to certain special economic zones (SEZs, see, e.g., Kleibert (2018)) or particular regions for foreign tourists only like those currently operating in Mongolia, East Timor and Laos. Their model has the potential to be further expanded nationally or adopted by other countries like Indonesia and Thailand, where casino gaming is still not legalized.

Given that the Chinese government will not approve casino gaming in Mainland China in the foreseeable future even though there might be many business opportunities associated with the potential demand for casino tourism in Asia as a whole, it is likely that investors could still be interested in countries that are currently in limbo (e.g., Taiwan and Thailand) even though jurisdictions with legalized gaming may continue to expand their business scale. In other words, casino tourism in the ESEA will continue to evolve and change at least in the next ten years.

1.2 Performance of the industry: an overview

To better understand the dynamics that attract investors who are interested in casino tourism, or in contrast, the increasing concerns of the different governments and local communities to the expansion of this industry in the ESEA, one should first refer to the changing business performances of this industry in the various destinations as reflected by the changes in their business turnover in terms of GGR (which also reflects the spending or gambling loss of tourists in enjoying

Table 1.1 Legal status of casino tourism in East and Southeast Asia in 2023

	Country/region	*Casino tourism*
East Asia	China	
	Mainland China	×
	Hong Kong SAR	×
	Macao SAR	✓
	Japan	•
	Mongolia	①Y
	North Korea	②Y
	South Korea	✓
	Taiwan (*de facto* country)	③N
Southeast Asia	Andaman & Nicobar Islands	×
	Brunei	×
	Cambodia	✓
	East Timor	②Y
	Indonesia	×
	Laos	②Y
	Malaysia	✓
	Myanmar	④Y
	Singapore	✓
	Thailand	×
	The Philippines	✓
	Vietnam	✓

Source: Compiled by the author with information derived from various online articles and reports current as of June 2023.

Notes:
✓ : casino gaming is legalized, operating license(s) issued and businesses operating
• : casino gaming is legalized, operating license(s) issued but businesses not operating yet
× : casino gaming is illegal
① : casino gaming is approved or not formally banned in certain hotels for foreigners only
② : casino gaming is allowed in the special economic zone (SEZ) and for foreigners only
③ : casino gaming is approved by the central government in offshore islands but not yet approved at the local level
④ : casino gaming was approved in 2019 but the necessary laws and regulations have yet to be promulgated. Nevertheless, some casinos have operated openly.
Y : casino(s) in operation
N : no casinos in operation

gaming activities). Their business performances in 2010 and 2019 are summarized in Table 1.2, respectively.

As shown in Table 1.2, the market size of the casino gaming industry in the ESEA in 2019 is 1.69 times that of 2010. By excluding the extraordinary performance (a large base) of Macao, the market size actually doubled (2.18 times). As for Macao, it is worth mentioning that in comparison to 2003 (with a reported GGR of USD3.58 billion) before the first foreign casino opened its doors (i.e., the Las Vegas-based Sands casino in 2004) when the monopoly structure of casino gaming was replaced with an oligopoly structure in 2002, the GGR of the industry indeed increased 9.20 times. As for the performance of some of the other jurisdictions, it is evident that despite the differences in their absolute market size, their relative

Table 1.2 Gross gaming revenue of major casino tourism destinations (2010 vs. 2019, billion USD)*

Destination	*GGR (2010)*	*GGR (2019)*	*% Change*
Macao SAR, China	23.543	36.557	55.3%
South Korea	1.965	2.515	28.0%
Cambodia	0.140	1.760	1,157.1%
Malaysia	0.948	2.514	165.2%
Singapore	2.800	3.350	19.6%
The Philippines	0.697	4.180	499.7%
Vietnam	0.069	0.107	55.1%
Total	30.162	50.983	69.0%
Las Vegas Strip	5.167	6.452	24.9%

Sources:
Macao: https://www.dicj.gov.mo/web/en/information/index.html
South Korea: 2010: Lee et al. (2018) (1 USD = 1129,9433 won)
2019: https://www.ngcc.go.kr/eng/stats/scale.do (1 USD = 1,165.29 won)
Cambodia: 2010: https://www.nagacorp.com/eng/ir/reports/ar2010.pdf
2019: https://igamingbusiness.com/casino-games/nagacorp-posts-full-year-profit-despite-covid-19-casino-closures/
Malaysia: 2010: https://www.pwc.com/gr/en/publications/assets/global-gaming-outlook-2011-2015.pdf
2019: https://www.ggrasia.com/gen-malaysia-revenue-to-2019-level-only-in-2022-nomura/
Singapore: 2010: https://www.casinonewsdaily.com/blog/2015/05/26/singapore-gross-gaming-revenue-analysis/
2019: https://www.ggrasia.com/singapore-casino-revenue-75pct-of-2019-in-2022-fitch/
The Philippines: 2010: https://www.pagcor.ph/annual-reports/annual-2010/pagcor-annual-report-2010.pdf
2019: https://www.pagcor.ph/annual-reports/annual-2019/pagcor-annual-report-2019.pdf
Vietnam: 2010: https://www.pwc.com/gr/en/publications/assets/global-gaming-outlook-2011-2015.pdf
2019: https://www.gamblinginsider.com/news/9853/local-gamblers-contribute-to-double-digit-revenue-growth-for-vietnam-casinos
Las Vegas Strip: https://gaming.library.unlv.edu/reports/NV_departments_historic.pdf

* Some of the numbers presented in this table may be a bit different from other sources owing to factors such as the exchange rate and different sources adopted. Nevertheless, the differences are small and would likely not affect the purpose of this table which is to show the changes and differences in the GGRs between the major casino tourism destinations in the ESEA. Also, considering the unexpected shocks from the COVID-19 pandemic on the tourism industries from 2020 to 2022, and the difference in the public and travel restrictions policies adopted by the different governments of the ESEA, the GGR data during this period of time might not reflect the real developments in these destinations. Thus, a comparison between 2010 and 2019 may offer a more realistic picture about their progress. The recovery of the various destinations and their prospects will be explicated in the following chapters.

performances are still striking (e.g., the relative scale of the industry in Cambodia and the Philippines increased remarkably). Other than Singapore, it is clear that the growth in GGR in the ESEA destinations is higher than that of Las Vegas (comprehensive analyses and discussions of the individual jurisdictions will be presented in the later chapters of this book).

In practice, although it is difficult to accurately pinpoint the contribution (or causal relationship) of casino gaming to the non-gaming spending of tourists in a destination (and *vice versa*), it is reasonable to think that there is a positive

Table 1.3 International tourism expenditure in major casino tourism destinations (2010 vs. 2019, billion USD)*

Destination	*ITE (2010)*	*ITE 2019*	*% Change*
Macao SAR, China	0.883	2.034	130.4%
South Korea	20.788	35.339	70.0%
Cambodia	0.131	1.162	787.0%
Malaysia	9.258	13.694	47.9%
Singapore	18.700	27.690†	48.1%
The Philippines	5.964	12.926	116.7%
Vietnam	1.470	6.460	339.5%

Sources: https://data.worldbank.org/indicator/ST.INT.XPND.CD

† https://www.budgetdirect.com.sg/travel-insurance/research/singapore-tourism-statistics (data for Singapore in 2019 are not included in the latest World Bank database)

* Tourism expenditure does not include spending on casino gaming. ITE denotes international tourism expenditure.

relationship between these two sectors which practitioners tend to also agree. For example, the international tourism expenditures shown in Table 1.3 can infer that the growth in the non-gaming tourism turnovers between 2010 and 2019 in the major casino tourism destinations and their GGRs are highly correlated. The correlation coefficient[1] between the changes in percentage of GGR (as presented in the last column of Table 1.2) and those in international tourism expenditure (as presented in the last column of Table 1.3) is 0.8372.

Given the aforementioned monetary performance and positive correlation between the growth in casino gaming revenue and tourist expenditure on non-gaming activities, it could be argued from a business standpoint that the successful development of casino tourism would have a higher economic value as opposed to the individual operation of the two sectors. Thus, the public decision to embed casino gaming into a tourism destination would generally be necessary to advance the overall performance of the industry and its economy. Nevertheless, monetary performance is clearly not the sole factor which gains public approval to legalize casino gaming or facilitate its continuous expansion in a tourism destination. To a large extent, the observed and changing performances of casino tourism in any destination may indeed be contingent on a number of social and political (including administrative) factors which include the possible adverse impacts of gambling on a society. In certain circumstances, the performance of this industry from a societal standpoint may include other factors that could be common in principle but different in practice owing to the different contextual settings across the different societies (which will be examined in the later chapters).

1.3 Research questions

Considering the progress, existing arrangements and performance of casino tourism in the ESEA, the practices and prospects of this industry in the region as a whole and in different individual jurisdictions clearly warrant in-depth investigations. To obtain realistic insights into this industry, the ten following questions need to be asked in the first place; it will help obtain the common grounds that can well clarify developing the frontiers of casino tourism in the ESEA. To a certain extent, although the responses to these questions may not be the same due to the different perspectives (e.g., the respective interests of business firms, regulators and different ethnic groups in a society on the practices of casino tourism will not be the same), addressing these questions in a meaningful way is important to anyone who would like to understand the factors that are crucial to ensuring the appropriate organization and hence sustained growth of this industry.

Ten questions to ask that will develop the frontiers of casino tourism:

1. What are the essence and attributes of casino tourism as a modern service industry?
2. What are the particular contextual settings and forces that frame the evolution and modern development of casino tourism in the ESEA?
3. What are the arguments (factors) for the development and expansion of casino tourism?
4. What are the arguments (factors) against the development and expansion of casino tourism?
5. What are the factors that result in similarities or differences in the behavior of the market participants and hence the business routine and performance between different jurisdictions?
6. What are the conditions for the sustained growth of casino tourism under the particular contextual settings of the ESEA?
7. What are the factors that may change the existing path of development of a casino tourism destination?
8. What are the impacts of social and political developments on the prospects of casino tourism?
9. What are the likely changes in the frontiers of casino tourism in the 2020s and 2030s?
10. Would the industrial structure and business organizations of casino tourism in various destinations exhibit convergence or divergence in the long run?

When asking these questions, the social, cultural, political and economic settings that may be specific to different societies are to be taken into consideration concurrently because these settings will have direct and indirect impacts on the practices and progress of casino tourism. Thus, the findings will be a valuable reference that can enhance the understanding of firms of the opportunities and threats associated with the progress of this industry in different societies. These will undoubtedly facilitate firms to formulate effective business strategies for the operation/management of their gaming business and combine non-gaming facilities that also align with social interests. In addition, answering these questions may provide realistic guidelines to policy makers in designing and implementing a quality regulatory system to reap the benefits from the practices of this industry and reduce its potential adverse impacts on society.

1.4 Approaches

To examine how the frontiers of casino tourism can be furthered to explain the progress and prospects of this industry in the ESEA, a pragmatic paradigm is formulated[2] (see Figure 1.2). It aims to provide a general framework for the analyses and arguments to be presented in this book. Unlike the analytical frameworks constructed to study other industries like agriculture, manufacturing, banking and e-commerce, a primary factor that needs to be considered for the practices of casino tourism is the legitimacy of casino gaming which requires government approval through the promulgation of gaming laws in the first place (as indicated by *a* in Figure 1.2). Subsequently, legally approved casino firms obtain the rights to

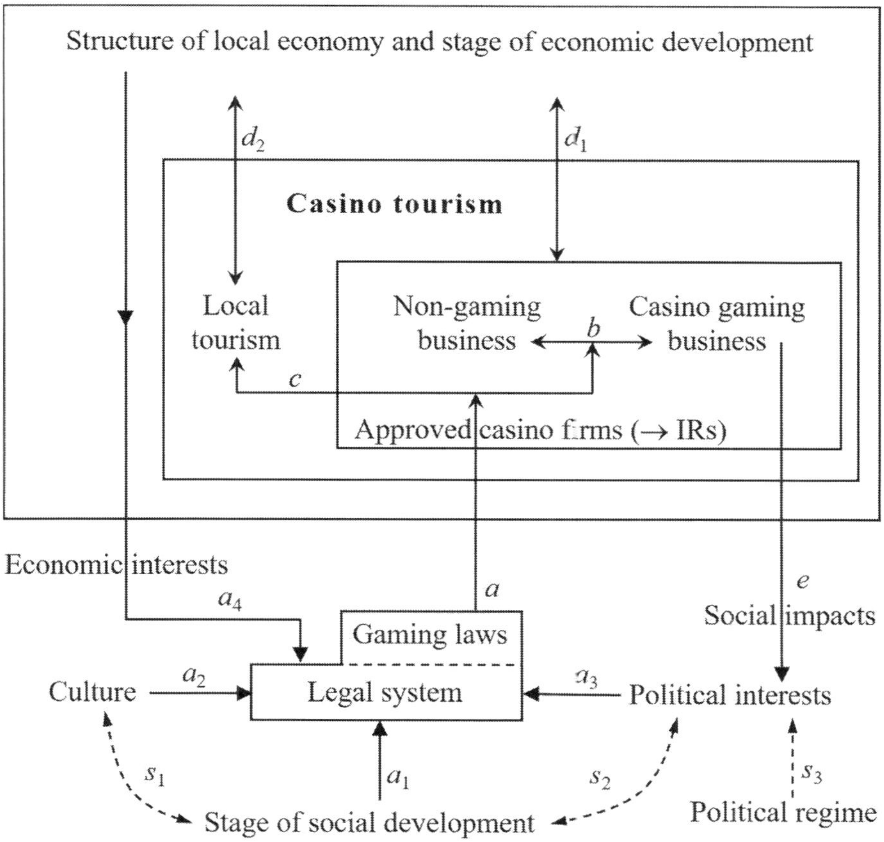

Figure 1.2 Analytical framework for furthering casino tourism

Source: Author

carry out gaming and non-gaming businesses simultaneously within their properties like the IRs (i.e., *b* in Figure 1.2). For any casino tourism destination, a number of interacting social, cultural, political and economic (SCPE) forces (as shown by a_1, a_2, a_3 and a_4, respectively, in Figure 1.2) that are specific to society clearly have crucial influence throughout the legislative process and the implementation of gaming laws although these laws are government designed and executed under its legal system.

Owing to the nature of casino gaming as a socially controversial industry, the design and contents of gaming laws for a destination and their implementation (including compliance) may not simply reflect a consequence (or trade-off) of the interactions between the economic interests (a_4 in Figure 1.2) pursued by different groups that potentially benefit them, like firms, investors and laborers, and the political interests of the government (a_3 in Figure 1.2), say, for tax revenue and political influences. Rather, social beliefs and values held by the populace under different stages of social development (a_1 in Figure 1.2), like the progression of the education and social protection systems, and culture (a_2 in Figure 1.2), like religion and faith, also have crucial impacts on the ways that casino gaming is to be organized and regulated under the laws. To any society, the stage of its social development is indeed interplaying with its culture (s_1 in Figure 1.2) and political interests (s_2 in Figure 1.2), while the political interests and structure of the government are largely shaped by the political regime which is specific to the society (s_3 in Figure 1.2).

According to the interplays of the SCPE forces which are reflected in the design and implementation of the gaming laws (i.e., through *a* in Figure 1.2), business practices and routines would then be formulated by the market participants (i.e., the practices of *b* and *c* as indicated in Figure 1.2), while the business scale and scope of casino gaming and its interrelationship with other non-gaming tourism sectors in a destination are restricted at the outset. As a modern industry, the practices of casino tourism unarguably interact with other local economic sectors (e.g., manufacturing, construction, wholesale, transportation, banking, schooling and training, the labor market) in the economy directly through the operations of the IRs, and indirectly through their connections with local tourism outside the IRs. These interactions are depicted in d_1 and d_2 in Figure 1.2. For example, the demand of IRs for labor will have different impacts (both positive and negative) like wage level and other employment conditions in the local labor markets for the tourism and non-tourism sectors. Consequently, feedback will be generated which may change the economic interests (i.e., changes in the force through a_4 to the legal system) for possible adjustments to the gaming laws.

Besides, social impacts under the practices of casino gaming, like increase in income and perceived improvement in material well-being on the one hand, and crime, family issues and questionable economic activities on the other hand, may produce crucial feedback (as shown by *e* in Figure 1.2) to the government so that it can consider any necessary actions to change the contents of existing gaming laws and their practices in the legal system. Given the interests and interactions between the various stakeholders in society (represented by s_1 and s_2 in Figure 1.2), the amendment could either expand the industry or limit its development (evidence will be presented in the following chapter to support this argument). Actually, due to the differences in the social and economic development stages, and culture and political regimes, evidence shows that the practices and social impacts of casino gaming differ across various casino tourism destinations in the ESEA.

Before proceeding to the discussions in the coming chapters, it is important to underline that even though gaming laws and their implementation in a legal system are specified as a unique factor at the outset in framing the further development of casino tourism, other factors beyond laws, like the direct impacts from the SCPE forces (which represent the contextual settings for market activities), technological advancement and innovation, the external environment, etc. on the actual performance of casino tourism, will also be included in the analyses. With reference to the aforementioned arguments in Figure 1.2, this paradigm can be used to examine the factors that further casino tourism in the various ESEA destinations by disclosing the SCPE forces that are common to the different societies and those that differ between them. With reference to Figure 1.2, the factors beyond the gaming laws which are associated with the contextual settings and technological progress across different destinations will also be revealed and examined.

1.5 Organization of this book

A detailed investigation and analysis of the development process associated with casino tourism in the ESEA are presented in this book based on the aforementioned overviews of the composition and progress of casino tourism in the ESEA, the questions raised and the approaches used to explicate the furthering of the frontiers of casino tourism. The book comprises the following chapters. In Chapter 2, the specific historical and contextual settings in ESEA that contribute to the evolution and modern development path of casino tourism in various countries/regions and the dynamics that contribute to their modern development are elucidated. These provide a realistic basis to further the studying and understanding of the similarities and differences in the progress and performance among the various casino tourism destinations and assess the factors that contribute to furthering their development, and hence their prospects.

In Chapter 3, the significance of social economics and political economy in academic studies is examined, and the development of casino tourism in a society is revealed. The impacts of the socio-economic factors (e.g., ethnicity, education, income) and government interventions (through the designs and enforcement of laws and regulations) on the practices of this industry in general and particularly in the ESEA as well as their interplays in a dynamic environment are scrutinized. Besides, reasons to advocate and oppose the expansion of casino tourism in a jurisdiction are analyzed and discussed.

To show the development of the frontiers of casino tourism in the ESEA, the evolution of Macao from a gambling hub in the mid-nineteenth century to a modern casino tourism destination in the twenty-first century provides an interesting case with a large volume of evidence. Thus, the distinctive contextual settings which have led to the development of this industry in Macao are presented in Chapter 4. Also, the enviable business performance in terms of the reported GGR and the tourism development of the city in the first two decades of the twenty-first century; dilemma associated with the practice of the "Macao model," which has led to high volatility and low business integrity issues; critical actions taken by the Macao SAR government in the beginning of the 2020s to radically change the organization of the industry; and the prospects of this industry under the new organization are scrutinized and analyzed in this chapter.

In contrast to the progress of Macao and other casino tourism destinations in the ESEA, Singapore is a late comer. Nevertheless, the success of the "Singapore model" reveals that a strictly and effectively regulated business environment can ensure the sustained growth of this socially controversial industry. In Chapter 5, the essence of the IR model as innovatively and strategically formulated by the Singapore government at the outset to unfold the development of casino tourism is explicated. Then, business practices and routines under a highly regulated system and the performance of the industry therefrom are elaborated. In a dynamic environment, potential impacts from the actions taken by the government to complete and renew the related laws and regulations over time on the continuous growth and development of the industry are discussed. Finally, the implications of the "Singapore model" on the ongoing development of casino tourism in the ESEA are also explored.

In Chapter 6, the specific SCPE forces that have pushed the frontiers of casino tourism in Malaysia and South Korea are examined respectively. Despite the common goal to reap the economic benefits, differences in the social and cultural contexts, as well as political and legal systems in these two countries, actually explain for the differences in their paths of development, and organization of the industry and its interplays with the local community and economy. In particular, the story of Genting Highlands, which gained the ultimate approval of the Malaysia government in 1969 to conduct casino tourism business exclusively in the country and the implications of its strategy to venture out and develop its IRs in the overseas markets, is told. On the other hand, the essence of the progress of this industry in South Korea from the establishment of foreigner-only casinos in the 1970s in some of the hotels to the development of a landmark casino resort located in a remote area in 2000 called Kangwon Land, which offers the only casino in the country that admits local citizens, is analyzed.

Unlike the other countries, casino tourism was initiated in the Philippines through the incorporation of a government-owned monopoly, the PAGCOR, in 1977. The particular forces that led to changes in the structure and organization of this industry in the Philippines and the reasons for the government to privatize PAGCOR in the mid-2020s are revealed in Chapter 7. In this chapter, the excitement and issues associated with the progress of casino tourism in four countries in the Indochina Peninsula (Cambodia, Vietnam, Laos, and Myanmar) of the SEA are also explicated. A wealth of evidence shows that while the particular geographical locations of these countries allow industry practitioners to pursue business opportunities for expansion purposes, drawbacks associated with the political and legal systems indeed inhibit the sustainability of their growth and real development. To this, the necessary improvements in their systems and prospects of their casino tourism are highlighted.

In Chapter 8, impacts from the critical social and political concerns about the existing gambling problems in Japanese society on the nearly two-decade-long process that the Japanese government went through to finally legalize casino gaming in 2018 are scrutinized. Under the particular requirements of the gaming laws, one of the three potential casino gaming licenses has been granted to an IR development project in Osaka in 2023. Despite the attraction and potential competitive edge of the project in Osaka, the first IR will only commence business at the earliest in 2030. To this, the prospects of this market in Japan will be evaluated. Then, the reasons for the back-and-forth debates on the legalization of casino gaming in Taiwan and endless wait for its development are discussed, although the establishment of casinos for tourism development in the offshore islands of Taiwan has been in principle approved by the central government in 2009. Finally, the potential path and industrial organization for casino tourism development in Thailand based on the actions taken by its government in the mid-2020s to legalize casino gaming are discussed. To this, the likely impacts on the frontiers of casino tourism in the neighboring countries of Thailand (i.e., Cambodia, Laos and Myanmar) are elaborated.

In Chapter 9, the prospects of casino tourism in the ESEA are further explored. To determine the prospects, influences from the changes in the "China factor," which significantly contributed to the rapid growth and expansion of casino tourism in the ESEA in the 2000s and 2010s, are examined. These include changes in the policies of the Chinese government and spending behavior of the younger mainland Chinese outbound tourists. Based on the facts and evidence presented and analyzed in this book, it is argued that to sustain the growth and development of casino tourism in the ESEA, industry practitioners have to introduce proactive and ongoing innovative changes. Besides, while the rapid changes in the mass market for casino tourism have been an irreversible trend, effective linkages between the non-gaming facilities of the IRs and local tourism resources in the various destinations are indeed indispensable to enhance their overall attraction (competitive edge). In conclusion, the importance of the role of a government at the outset to develop the frontiers that would effectively sustain the growth and development of this socially controversial industry is revisited.

Notes

1 Correlation coefficient is a statistical measurement for the linear relationship between two quantitative variables. The value of the correlation coefficient falls between −1 and +1. While +1 indicates a perfect positive linear relationship (i.e., when the quantity of one variable increases, the value of another variable also increases, and *vice versa*), −1 indicates a perfect negative relationship, and 0 implies no relationship between these two variables. However, correlation does not imply causation. Related calculations and discussions can be found in any basic statistics textbook.

2 This paradigm represents a re-consideration/-construction from my previous studies on casino gaming and casino tourism in the case of Macao and Asia. See Figure 6 in Siu (2013), Figure 1 in Siu (2016) and Figure 1 in Siu (2023).

References

Ashton, Harry and Laurie Korpi. 2008. "Asia Awakens: The Growth of Casino Tourism." Gambling Compliance Ltd. Available at https://docplayer.net/15330577-Page-1-gamblingcompliance-asia-awakens-the-growth-of-casino-tourism-gamblingcompliance-research-team.html. Accessed June 15, 2021.

Barta, Patrick. 2012. "Cambodia Bets Casinos Can Boost Tourism." (April 3, 2012). Available at https://www.wsj.com/articles/SB10001424052702304750404577317094266574870. Accessed June 14, 2021.

Chan, Viviana. 2023. "Osaka IR Opening Delayed to Autumn 2030, Costs Rise $1.3bln." *Asia Gaming Brief* (September 5, 2023). Available at https://agbrief.com/news/japan/05/09/2023/osaka-ir-opening-delayed-to-autumn-2030-costs-rise/. Accessed September 6, 2023.

China Daily. 2005. "Singapore Gives Green Light to Two Casinos." (April 19, 2005). Available at https://www.chinadaily.com.cn/english/doc/2005-04/19/content_435463.htm. Accessed July–April 22, 2005.

Dolezal, Claudia, Alexander Trupp and T. Bui Huong. (eds). 2020. *Tourism and Development in Southeast Asia*. New York, NY: Routledge.

Foster, Tony and Nguyen Tam Dan. 2022. "Casino Investment in Vietnam." (June 27, 2022). Available at https://www.freshfields.com/en-gb/our-thinking/knowledge/briefing/2022/06/casino-investment-in-vietnam/. Accessed August 14, 2022.

Gunesch, Konrad. 2017. "Socio-Economic, Cultural and Ecological Implications of Gaming and Tourism in Asia: The Pacific Region between Tradition and Modernity Regarding Comprehensively Entertaining, Family-Oriented and Culturally Preserving Heritage Travel." Available at https://doi.org/10.19275/RSEPCONFERENCES103

Kleibert, Jana M. 2018. "Exclusive Development(s): Special Economic Zones and Enclave Urbanism in the Philippines." *Critical Sociology* 44 (3): 471–485.

Lee, Donghwa, Yen-Soon Kim, Eunha Myung and Hyun Kyung Chatfield. 2018. "Integrated Casino Resort Development in South Korea: Perspectives from the Government Representatives and Industry Professionals." *Journal of Tourism &* Hospitality 7: 375. https://doi.org/10.4172/2167–0269.1000375.

Lee, Young Gi. 2017. "The 'Singapore Model' in Gaming: Applications in South Korea." *UNLV Theses, Dissertations, Professional Papers, and* Capstones: 3203. http://dx.doi.org/10.34917/13444550.

McCartney, Glenn J. 2016. "Play the Cards, Roll the Dice – The Integration of Casinos Within Asian Tourism Destination Development Strategy." *UNLV Gaming Research &* Review Journal 20 (2): 119–140.

O'Connor, Devin. 2021. "Japan Integrated Resort Bill Becomes Law July 19, Bidding Begins in October." (July 15, 2021). Available at https://www.casino.org/news/japan-integrated-resort-casino-bill-becomes-law-july-19/. Accessed July 18, 2021.

Siu, Ricardo C. S. 2013. "The Economics of Asian Casino Gaming and Gambling." In *The Oxford Handbook of the Economics of Gambling*. Edited by Vaughan Williams Leighton and Donald S. Siegel, 55–86. New York, NY: Oxford University Press.

Siu, Ricardo C. S. 2016. "Path Dependence in the Industrial Development of Casino Gaming: The Lesson from Macao." *Manuscript* (presented in the 16th International Conference for Gambling and Risk Taking, in Las Vegas, USA, June 6–10, 2016).

Siu, Ricardo C. S. 2023. "Back to the Future: Constructing Macao as a World-Class Casino Tourism Destination under New Gaming Laws." Gaming Law Review: Economics, *Regulation, Compliance, and Policy* 27 (7): 326–342.

Zhang, Juan. 2017. "Introduction: Integrated Mega-Casinos and Speculative Urbanism in Southeast Asia." *Pacific Affairs* 90 (4): 651–674.

Zhu, Michael. 2022. "The Philippine Casino Gaming Market." (December 1, 2022). Available at https://ggbmagazine.com/article/the-philippine-casino-gaming-market/. Accessed February 21, 2023.

Additional readings

GGRAsia. 2019. "Mongolia Bans Civil Servants from Gambling in Casinos." (April 4, 2019). Available at https://www.ggrasia.com/mongolia-bans-civil-servants-from-gambling-in-casinos/. Accessed August 6, 2019.

McCurry, Justin. 2023. "Japan Approves Building of First Casino." (April 14, 2023). Available at https://www.theguardian.com/world/2023/apr/14/japan-approves-building-of-first-casino. Accessed May 2, 2023.

Shin, Hyonhee. 2017. "North Korea Is So Desperate for Hard Currency That It's Legalized Racetrack Gambling." (October 17, 2017). Available at https://time.com/4985283/north-korea-racetrack-gambling-horse-racing/. Accessed July 14, 2021.

Sims, Kearrin. 2017. "Gambling on the Future: Casino Enclaves, Development, and Poverty Alleviation in Laos." *Pacific Affairs* 90 (4): 675–679.

2 Evolution and modern development of casino tourism in East and Southeast Asia

A comparison of the history of casino gaming in the Western context, i.e., Monte Carlo in Monaco (Western Europe) and Las Vegas in the United States of America (North America), shows that the origins of gaming as a business and its evolution to casino tourism as a modern industry in the ESEA exhibit clear differences in terms of the business organization and regulatory systems, and thus the practices and path of development of the industry. In fact, the differences are first reflected in the usage of the word "casino" in the Western societies and "gambling den" in the ESEA to refer to the characteristics and practices of the gaming business. The *Oxford Dictionary*, for example, defines "den" as an idiom which describes "a place where evil or immoral activities go on."

In contrast, the word "casino" originated from the Italian word *casa*, which translates into a "house" that is "used for social amusement" (according to the 10th edition of the *Merriam Webster's Collegiate Dictionary*). Also, Thompson (2015, 43) indicates that in other dictionaries, casinos are defined as "Italian summer villas," "brothels," or "social clubs." It can be seen from these definitions that despite the social concerns, casino gaming and the associated business activities in the ESEA are evolving from a more socially controversial definition and status than in the Western societies. Thus, to understand the frontiers and further the progress and prospects of this industry in the ESEA, the specific contextual settings in the various countries/regions and the impacts of these settings on the evolution of their casino tourism endeavors need to be first explicated.

In this chapter, some of the distinctive features associated with the practices and evolution of casino tourism in the Western societies are first described. These provide the essential evidence used to compare and contrast the paths of development and actual performances of the various jurisdictions in the ESEA. As indicated in Chapter 1, a set of SCPE (i.e., social, cultural, political and economic) forces has to be elucidated to obtain a realistic understanding of the frontiers for developing this industry. Thus, the particular contextual settings in the ESEA and various countries/regions and their interrelationships with the evolution of gambling activities historically and casino tourism from the 1960s to the 1990s are scrutinized and revealed in the second and third sections. Then, the factors which have contributed to the dynamics for the modern and rapid development of this industry up to 2023 are explicated in the fourth section. In addition to the facts and trends that are

DOI: 10.4324/9781003279884-2

identified, the prospects of this industry and projected changes are examined in the last section.

Takeaways from this chapter:

1. Casino tourism is a socially controversial industry owing to the moral issues and social concerns associated with the practices of casino gaming.
2. The culture and stage of social development of countries/regions in the ESEA clearly differ.
3. The political and legal foundations of most casino tourism destinations in the ESEA are greatly influenced by their respective historical relationships with the Western world, especially Western Europe.
4. The practices of the legal systems (in terms of the design and implementation of laws and regulations) in many countries/regions in the ESEA are mainly determined by the political interests of the ruling parties and the changes in their interests.
5. The evolution/progress of casino gaming in most countries/regions in the ESEA shows particular interplays between law and customs to achieve various economic benefits and obtain political interests in the first place.
6. The modern development of casino tourism in the ESEA in the twenty-first century is first promoted by the dynamics resultant of the decision of the Macao SAR government to eliminate the gaming monopoly in 2002 by adding two world-renowned mega-casino resorts and then pushed forward by the decision of the Singapore government to legalize casino gaming in 2005 as a formal business segment to be included in the development of their IRs.
7. While demand for casino tourism by the Chinese is a primary force that contributes to the rapid growth of this industry in the ESEA in the 2000s and 2010s, the origins of the patrons and their preference for casino tourism are changing over time.
8. The industrial organization and business routines of casino tourism in the ESEA are moving toward a more modernized path of development. While the non-gaming tourism sector continues to expand, the gaming sector will assume more social responsibility.

2.1 Characteristics and a short history of casino tourism

The practices of casino tourism can be conceptualized as a particular category of the tourism industry in which casino gaming services are provided to visitors as either a primary or a complementary (or an alternative) form of entertainment. Historical global evidence usually shows that while some casino tourism destinations have evolved from districts with certain types of casino gaming business initially and non-gaming attractions and hospitality capacity following by extension in their path toward modern development, other destinations have subsequently legalized casino gaming in their attempts to revitalize their tourism sectors and local economies. Nevertheless, these two trajectories in the development of casino tourism are not always a smooth and linear progression in practice because in some destinations, some form or shape of casino gaming and casino tourism activities may have already been in place informally prior to their modern development in the twenty-first century.

2.1.1 Experiences of Western societies

The development of casino tourism in the Western societies in general has been a process in which legalized casino gaming activities are commercially operated and supplied in a restricted destination. The businesses meet and stimulate the anticipated demand by offering a wide range of non-gaming hospitality and leisure facilities/services simultaneously for visitors to enjoy during their trips away from home. For example, in 1806 when Napoleon first legalized casinos in France as a measure to "close down the popular gambling rooms" (Kingma 2008, 31) in the country, these legalized venues were designated in "Paris and some watering places . . . serving the interests of tourism and the state" (ibid.). In addition, casino gaming has been approved and introduced as a particular form of entertainment to enhance the attractions and financial turnovers of the industry, which is also a strategy to promote the urban development of the destinations.

In addition, Franke (2022) states that after the first three decades that Monte Carlo was legalized as a casino district in 1866, "it became a cosmopolitan pleasurescape designed for nonlocal, seasonal population of gamblers and consumers, separated from the neighboring urban entities" (Franke 2022, 1247). This is because "[f]rom the very beginning, the government and casino executive stress that a new gambling venture in Monaco would have to include wide-ranging urbanization projects" (ibid., 1249) to enhance the infrastructure and hence the efficiency of the city as a tourism destination. Indeed, the effective alignment of casino gaming with non-gaming facilities, like catering, shows and other forms of entertainment did concur with the original public interests and planning for the long-term development of Monte Carlo, and hence contributed to a sustained path of development in the twentieth century.

As for Las Vegas, its particular geographical location before its formal establishment in 1905 as "a place where travelers stopped on their way to somewhere else . . . which became a welcome stopping-point for trade caravans" (Schwartz 2003, 26) provided opportunities for firms to introduce gaming as a form of entertainment service (together with other services like bars and pools) for travelers while they were enjoying their stay. Nevertheless, it is also a well-known fact that before the 1960s, casino gaming was largely influenced by organized crime (see, e.g., Richard (2018)) so Las Vegas was labeled as a sin city for decades in the first half of the twentieth century. Despite the negative social image, however, a crucial factor led to the modern development of Las Vegas; in 1969, the Nevada state government promulgated the Corporate Gaming Act (Richard 2018; Eadington 1999), which allowed publicly listed companies to develop (mega) casino resorts under strict supervision. Thereafter, a new "legal climate for American casinos" (Eadington 1999, 175) ensued and Americans considered "the casino" "not a sybaritic den of vice, but a place to work or go for entertainment: to see a movie, play bingo, hit the buffet, or gamble" (Schwartz 2003, 1).

2.1.2 *Unique features of East and Southeast Asia*

Casino-type games in the ESEA like dice games and Chinese Fantan (see, e.g., Paulès (2010) for a description of the operation and practices of this game and related issues) in the nineteenth and early twentieth centuries were largely operated in gambling dens by individual operators in an almost lawless jurisdiction. For example, a common scenario which depicts the Fantan game carried out in a gambling den in the early twentieth century is shown in Figure 2.1. In the image, there is a poster with four Chinese characters "dà shā sān fang," literally translated as "killing the three parties" or loosely translated as "winner takes all." These four characters convey a unique message to the patrons in the gambling dens: one of the individuals sitting at the table is particularly powerful and outwits the other three individuals so that they do not have the ability to "fight back." This is called "killing the three parties." So, it is evident that the gambling dens were largely venues for individuals to show off their gambling skills and satisfy their speculative mentality for monetary return instead of entertainment. Also, it is evident that non-gaming hospitality services were almost non-existent in the gambling dens.

Considering the nature of the gambling dens and their associated activities, their practices might not have any real relationship with tourism (neither within nor outside the gambling properties), not to say the contributions to urban development as compared to the history of casino gaming in the Western societies. In addition, the underdeveloped social environment and legal system in most of the countries/regions in the ESEA before the mid-twentieth century meant that gaming businesses in their various forms were largely operating in a lawless vacuum or poorly

Figure 2.1 Chinese fantan game in a gambling den (early twentieth century)

Source: https://www.lookandlearn.com/history-images/YN031407/Chinese-Fantan-game

supervised jurisdiction. Thus, it is a matter of fact that adverse social impacts like crime, scams, family problems etc. were commonly observed by the populaces in association with the practices and expansion of the casino-type gaming activities and businesses. The lack of opportunity in the social structure and unstable political environment, and hence the lagged social development in most of the ESEA countries/regions before the mid-twentieth century, meant that casino-type gaming activities/business were largely loosely organized and under-regulated, and influenced by the conventional ways of doing business (i.e., part of the culture) in the different communities (see, e.g., Price (1972)).

Indeed, in the development of casino tourism as a modern industry in the ESEA, the 1960s could be considered as a watershed era. First, although casino-type gaming businesses were already approved (legalized by the Macao-Portuguese government in the mid-nineteenth century), the first gaming law was promulgated only in 1961. It was the first law that explicitly stated that casino gaming license holders were to contribute to the economic development of the city by developing tourism. Following capital investment and construction works in the 1960s, the development of this industry in Macao since the 1970s well exhibit the evolution of Macao from a largely casino-based jurisdiction to an initial casino tourism destination model (synopses of this case will be highlighted in the following sections, and the particular role and dynamics of Macao in the modern development of casino tourism in the ESEA from 1990s onward will be explicated in Chapter 4). Other than Macao, another distinct and interesting case which contributes to the development of casino tourism in the ESEA is the approval of a casino gaming license by the Malaysian government in 1969 as part of a larger tourism development project in a particular region in the country (which will be discussed in Chapter 6). Unlike Macao, casino tourism in Malaysia did not evolve from a history of casino gaming.

As for the progress of the other casino tourism destinations from the 1970s to the 1990s (e.g., South Korea and the Philippines since the 1970s, and Cambodia, Vietnam and Laos since the 1990s) and the 2000s (i.e., Singapore), their respective paths of development and frontiers in the development of casino gaming are not the same, and even quite different from one another. To a large extent, the variations are attributed to the differences in their contextual settings.

2.2 Contextual settings of East and Southeast Asia

Evidence shows that the structure, organization and business routines of an industry in practice and its performance are not simply determined by economic factors like demand and supply in the markets and the economic activities of the governments. Rather, the contextual settings of a society like its geographical location, cultural attributes, social structure, political interests, legal system and level of technological development will have significant impacts on the practices and performances of an industry. In socially controversial industries like casino tourism, changes in political interests may have one-time or radical impacts which alter their path of development and the existing industrial structure and organization, while social development has continuous impacts on their practices. By the same token, the practices of an industry in a dynamic environment will also have feedback effects and interplay with the contextual settings like social development, public interests, and technological advancements in society.

2.2.1 Geographical location, cultural attributes and social structure

The different geographical locations and historical progress of the ESEA countries/regions mean that their cultural attributes and social structure are clearly different from those in other regions. Moreover, the cultural attributes and social structure of the ESEA countries/regions can be quite diverse. In general, some of the major studies on culture and religion indicate that the Chinese who historically migrated to different regions of the ESEA like modern Japan, Korea, Malaysia, Singapore etc. clearly impacted the social and economic activities of these communities (see, e.g., Weber (1951) and Wilkinson (1996)), which definitely included gambling. Nevertheless, the local culture (including religion) and historical progress of those societies also exhibit and retain their respective attributes. For example, although the Chinese are the second-largest ethnic group in Malaysia, Islam retains a dominant role in the Malay culture. The Philippines has also incorporated Islamic law into its legal system which only applies to Muslims.

On the other hand, Price (1972) claims that the intensity and types of gambling games played by the Chinese in north versus south China and the impacts on the various regions in Asia are different, although it is commonly claimed that gambling is a Chinese convention (see, e.g., Chien and Hsu (2006)). In general, evidence shows that "[i]ntensive gambling seems to be associated with intensive agriculture" (Price 1972, 176) which explains for the greater inclination of the Chinese in East Asia (mainly in south China and southeast Asia) to gamble rather than those in the northern part of Asia and West Asia. Also, Price (ibid.) finds that "[g]ambling is associated with the sedentary adult games of East Asia, rather than the more physical entertainment of North and West Asia."

Taoism and Confucianism have been historically the guiding social philosophies in Chinese society, and their interactions with Buddhism not only preserved the feudal social structure in China for over two and a half thousand years from the 1050s BC to the early 1910s AD but also influenced other societies in the ESEA. Thus, despite the technological advancements of ancient China (mainly represented by the four great inventions which included papermaking, printing, gunpowder and the compass) and its influences globally, the social structure and progress of most countries in the ESEA were largely stagnant, say, in terms of the low education levels and literacy rates before the mid-twentieth century. To a large extent, the everyday lives of the general public adhered to a set of traditional routines (social norms), while the economies in the ESEA were predominantly agricultural and handicraft-based until the mid-twentieth century.[1] These also included the social values and social orders to maintain relationships between the various social groups, especially the rulers and the ruled/governed.

Owing to the infiltration of Western cultures in the seventeenth century (e.g., Spanish, Portuguese, French and British occupancy of land in the ESEA countries), it is commonly observed that the modern culture and social structure of many countries in the ESEA are actually quite diverse with a fusion of varying degrees of their own respective culture, the Chinese culture and the different Western cultures. Indeed, the diversity in culture and social structure among the ESEA countries has increased since the mid-twentieth century. A major factor that has led to these changes is the active involvement/influence of Western governments and business firms to various extents and scopes, and in the different countries in the ESEA after the Second World War.

2.2.2 Political regimes and legal systems

Political regimes[2]

In contrast to the Western societies, the political regimes in the ESEA countries are quite diverse owing to their individual social settings and historical progress, as well as the influences from the Western world in the seventeenth century. In general, "[c]onstitutional monarchies, absolute monarchies, one-party states, federal states, liberal democracies, and military dictatorships are all present in the region" (Barkema et al. 2015, 461). This diversity was also well reflected in the ESEA countries where casino tourism is legalized. For example, although Malaysia, Singapore, the Philippines and South Korea are in principle considered to be democracies, their practices are not the same. Malaysia has a parliamentary democracy government with a constitutional monarchy, Singapore is a parliamentary representative democratic republic, the Philippines is more a democracy in general, and South Korea is a democratic republic.

The political regimes of Vietnam, Laos and North Korea are dominated by a communist one-party system: that is, the Communist Party of Vietnam (CPV), Lao People's Revolutionary Party (LPRP) and Workers' Party of Korea (WPK), respectively. Nevertheless, owing to the different political interests of the controlling parties, the organization and practices of the respective political regimes are not the same. On the other hand, the political system of Cambodia is still dominated by a communist one-party system (i.e., the Cambodian People's Party (CPP)) even though communism under the Khmer Rouge was formally removed in Cambodia in 1979 and the establishment of some individual political parties has been allowed.

As for the most dynamic casino tourism destination in the world, Macao is a non-sovereign city that was occupied by the Portuguese for hundreds of years until the end of 1999, when it was returned to China and formally established as an SAR of China under the "One-Country, Two-System" policy. Basically, the political regime in Macao is a multi-party executive-led system. Lastly, the political regime of Japan is a dominant-party bicameral parliamentary constitutional monarchy.

Legal systems[3]

Aside from their political regimes, the legal systems of the casino tourism destinations in the ESEA are also diverse. The diversity is due to the early influences of the colonial administrators of Portugal, France and Britain, and the active involvement of the US government in the modern development of the ESEA since the 1950s. Thus, the legal system of many of the ESEA countries is actually based on a civil or common law system. For example, the legal system of Macao is based on the Portuguese civil law system. Vietnam, Laos and Cambodia were occupied by the French before the mid-1950s, so the French civil law system comprises the foundation of their legal systems. Nevertheless, the advancement of their respective systems is also influenced by other internal and external historical forces. For instance, while the legal system of Vietnam is influenced by socialism's legal theory of the former Soviet Union, the legal framework of Cambodia has evolved into a hybrid system today, with influence from the Vietnamese legal system as a result of the 1979 Vietnamese invasion, as well as influence from the United Nations, which was involved in their legal affairs in the beginning of the 1990s.

On the other hand, Malaysia and Singapore, which were both historically British colonies, follow the common law system. However, Malaysia is an Islamic country, so Islamic laws are also found in its legal system. Both the legal systems of the Philippines and Japan are influenced by the common law system (Anglo-American laws) of the United States as they have a close relationship with the Americans. However, Islamic law is also incorporated into the legal system of the Philippines which applies to only Muslims as Islam is the second-largest religion in the country. Japan has a more complex system, which incorporates six legal codes based mostly on that of Germany, but also on the models of France and Switzerland, as well as the Japanese context. Japan can therefore be considered to have a hybrid legal system.

In considering the aforementioned diverse cultural attributes, social structure, political regimes and legal systems in the ESEA,[4] it is then reasonable to infer that the industrial structure and organization of casino tourism in other societies would also have evolved diversely. Based on these initial facts, the different paths of development and performances among the various casino tourism destinations in the ESEA and their possible progress in the foreseeable future will first be outlined in the following sections and then further explicated in detail in the following chapters.

2.3 Basis, organization and structure of the industry

Under the particular contextual settings (especially the level of social development and political structure) of the ESEA countries, casino tourism before the mid-twentieth century was usually an informal business sector which was predominantly the provision of gambling services by some of the power groups (including triad gangsters) to the community. To a large extent, gambling operators were carrying out their business in a loosely regulated or lawless jurisdiction, and their patrons were mainly visitors who resided within a short distance of the respective countries or from adjacent regions/countries. Besides, although some simple forms of accommodations, catering and entertainment services were provided to the patrons, these services basically catered to the gambling business. In other words, the organization of casino tourism (which was an extension of gambling dens and gambling houses) in the ESEA was predominantly a sort of gambling tour in nature between the mid-1800s and mid-1900s.

2.3.1 Coupling of tourism and casino gaming

Following the increasing political and economic interests of the Western world in the ESEA in the post–Second World War period (in the 1950s and 1960s) and the expansion of foreign investment and international trade in the different countries (except for Mainland China before 1978 when the country had closed-door social and economic policies with the Western world), industrialization and economic development in the various countries/regions led to the modern development of their societies and economies.[5] During the process, their governments widely supported the development of tourism as a modern industry which would attract revenue from international tourists and therefore directly bring in foreign exchange. Meanwhile, when the income of local households rose as their economy developed, so did their demand for domestic and cross-border tourism. Thus, the increased spending of international and domestic tourists drove the development of tourism in the ESEA.

As a business segment that can be incorporated into further tourism, casino gaming began to be formally incorporated as part of the entertainment options in the properties of some of the modern hotels and resorts for their visitors at the end of the 1960s. In contrast with the traditional gambling dens and gambling houses, casino gaming in these modern hotels and resorts was legalized under the gaming laws in the respective countries/regions. In general, the contents of these laws reflect the public interests in the different countries/regions through how casino gaming was organized, which also shows the scale and scope of this business segment and their connections with the local tourism sector.

For example, following the establishment of the Korean Tourism Organization (KTO) by the South Korean government in 1962 to promote the country to foreign visitors, the first legalized casino hotel, the Incheon Olympos Hotel Casino, was formally approved and opened in 1967. Nevertheless, casino gaming was restricted to foreign tourists only (Lee 2011, 451). The political interests which prohibited locals from participating in casino gaming meant that casino tourism in South Korea was available through small-scale casinos within approved hotels or resorts that were meant for foreigners only between the 1970s and 1990s. In 2000, a new casino resort called the Kangwon Land Casino was authorized as the only property in South Korea to admit local visitors for casino gaming activities. To a large extent, the decision of the South Korean government to allow locals to enjoy casino tourism represents public interest and a chance to capture this market as opposed to the spending of the locals on gaming outside the country if casino tourism was not available to them in South Korea.

2.3.2 Industrial structures

Before the twenty-first century, the evolution of casino tourism in other destinations in the ESEA like Vietnam, Cambodia and Laos usually went through a similar legalization process of casino gaming by their governments, and the related gaming facilities were approved in various individual hotels or small-scale resorts in the tourism districts. Besides, their expansion was generally initiated by the increasing number of government-granted licenses. Nevertheless, different types of industrial structures and organizations are also found in other jurisdictions in accordance with their respective political interests and legal systems. For example, the casino gaming business in the Philippines was dominated and operated by a state-owned corporation called the Philippine Amusement and Gaming Corporation (PAGCOR) from the 1970s to the mid-2010s; see, e.g., Lopez (2014). Under this structure, some of the casinos in the Philippines are found in privately owned hotels or resorts; however, most are operated directly by the PAGCOR in its own entertainment or hotel properties. In the 2010s, some of the privately owned casino/IRs also received approval to operate their own casino gaming. Considering the success of these privately owned IRs, the Philippines government decided to privatize the PAGCOR in the mid-2020s to improve the organization and long-term competitiveness of this industry.

On the other hand, the development of casino tourism in Macao and in Malaysia since the beginning of the 1970s are two distinct and interesting cases. In these two destinations, casino gaming was first approved by their government under a monopoly structure and entrepreneurship clearly played a particular role in influencing the evolution of the industry (see, e.g., Wong and Siu (2015) and Firzanah (2021)), but their practices had very different paths of development and generated different feedback effects from their respective societies and economies. Actually, the differences in their degree of social and economic development and legal systems are largely responsible for the differences in their business organization (details of these two cases will be explicated, compared and contrasted in the later chapters).

To recap, as a modern business segment for casino tourism, the organization and structure of casino gaming practices in the different ESEA destinations and their interplay with other local tourism sectors are determined by three major scenarios at the outset which include:

1. a limited but not fixed number of individual licenses;
2. a monopoly or oligopoly license; and
3. the co-existence of state-owned and privately-owned licenses.

In practice, while the preference of a government in these scenarios reflects its political inclination and interests in the role of this industry in its economy, this preference also reflects its considerations of the particular contextual settings of its society. Thus, this explains the different paths of the development of casino tourism in the ESEA countries/regions.

2.4 Landmarks for modern development and their dynamics

Despite the aforementioned contextual factors (mainly social and economic development and changes in political inclinations in the legalization of casino gaming) which have led to the differences in the industrial structure and organization, and timeline in the advancement of various casino tourism destinations in the ESEA, the spread of the Western-style casino hotel/resort business model in the 1960s as part of the effect and process of globalization (or Westernization) post–Second World War clearly constitutes as an essential dynamic for their development. In this process, public interests to institute a closer relationship between casino gaming and tourism in the various destinations as a measure to promote tourism and economic development are evident. In addition, gaming laws and regulations are also legislated according to the legal systems in the ESEA countries/regions to supervise the practices of this industry.

2.4.1 Landmarks of modern development

Using the conceptualized characteristics of casino tourism in the first section of this chapter, the landmarks in the early and modern development of this industry in the various destinations of the ESEA are listed in Table 2.1. In principle, these landmarks are exemplified by the opening of the first casino hotel (or casino resort or IR) in the respective destinations in which casino gaming has been explicitly approved by the gaming laws (i.e., *a* in Figure 1.2 of Chapter 1). In these properties, a series of hospitality and entertainment/leisure services is provided to the tourists and locals regardless of the fact that they spend money on casino gaming or not. In addition, their practices may also interact with other local tourism sectors outside the properties (e.g., hotels, restaurants and retail stores which are located in proximity to the casino hotels and resorts) and facilitate their development to various degrees.

Actually, Table 2.1 shows that the modern development of casino tourism in the ESEA has mainly proceeded through three phases. First, there is the establishment of a prototype casino model in a few of the destinations (South Korea, Macao, Malaysia, and the Philippines) in the 1970s which had similar configurations as

Table 2.1 First casino hotel/resort in ESEA casino tourism destinations

Country/region	*Casino name/Property name*	*Year of opening*
South Korea	Olympos Casino/Incheon Olympos Hotel Casino	1967
Macao	Casino Lisboa/Hotel Lisboa	1970
Malaysia	Genting Casino/Highlands Hotel (now Theme Park Hotel)	1971
The Philippines	The Philippines Village Hotel	1979
Vietnam	Do Son Casino	1995
Cambodia	Naga Casino/NagaWorld	1995
North Korea	Pyongyang Casino/Yanggakdo International Hotel	1995
Laos	Dansavanh Casino Resort	1996
Singapore	Marina Bay Sands Casino/Marina Bay Sands Singapore	2009
Japan	MGM	2030*

Source: Author's online searches with cross-checking.

* *Estimated*

those of Monte Carlo and Las Vegas. Second, some of the countries (Vietnam, Cambodia, North Korea and Laos) then launched their own casino hotels/resorts in the 1990s which "borrowed" elements from the casino tourism model of the players in the first phase. These borrowed elements included not only the infrastructure such as the general fixtures and facilities of casino hotels/resorts but also their marketing strategies to obtain the potential revenue from patrons in neighboring countries where casino gaming is outlawed. Third, there is the transformation from traditional and relatively small-/medium-scale casino hotels/resorts to mega/IRs in the first two decades of the twenty-first century under new dynamics based on Macao's restructuring of its industrial structure which was further accelerated by a new competitor, Singapore. Actually, transformation in the twenty-first century also involves significant revisions of the business models and business routines by the existing destinations and their increasing capital investment in non-gaming facilities.

2.4.2 Dynamics in the twenty-first century

While the development of casino tourism in the ESEA between the end of the 1960s and the end of the 1990s (i.e., the first two phases as indicated in the last paragraph) represented the startup and initial growth periods, most destinations experienced unforeseen and dramatic expansions from 2000 to 2019. Actually, three particular dynamics are found that contributed to these changes. First, the success of the Chinese economy generated considerable income and wealth effects from the demand side which drove the growth of tourism in general and casino tourism in particular globally.[6] Second, the establishment of Macao as an SAR of China in 1999 allowed the city to gain full support from the Chinese government under its "One-Country, Two-System" policy and take the opportunity to expand its casino tourism and become the largest casino destination worldwide in 2006 based on its GGR. Third, a critical decision was made by the Singapore government in 2005 to legalize casino gaming and formally (i.e., in a legal document) use the title "integrated resort" to describe the business segment of casino gaming.

In light of these new dynamics, the American Gaming Association conducted a survey in 2008 and found that "Asia may surpass the USA in casino gambling revenue by 2012, with growth driven by Macau, Singapore, and sites currently being considered in other Asian countries" (Kale and Spence 2009, 321). Indeed, this prediction came true. In the face of the increasing and potential demand (from both domestic and international markets) for casino tourism and the success of the new model for the development of large-scale IRs that are operating in Macao and Singapore, it is evident that other casino tourism destinations started to follow the same path of development since the mid-2000s. Supported by the changing government policies, new IRs were constructed and their doors opened in the 2010s in some destinations like the Resorts World Manila (which was renamed Newport World Resorts in 2022) in the Philippines, while some of the existing mega casino resorts like Kangwon Land in South Korea also expanded their facilities (especially non-gaming attractions) over time.

On top of ensuring and strengthening the allure and hence the competitive edge of their own industries in the ESEA, casino tourism development has also been clearly adopted by the different governments as a measure to foster urban development in the related districts despite "the fraught nature of urban (and peri-urban) processes" (Zhang 2017, 654) from the standpoint of social development. To this, it is clear that under the new dynamics and attraction of IR development, advocates generally believe that a net benefit could be derived through the modern development of casino tourism in the destinations (the disputes and downsides associated with the rapid expansion of casino tourism will be explored in Chapter 3, and some individual cases will be presented in the other chapters).

2.5 Existing structure/organization and probable changes

Following the rapid progression of casino tourism from 2000 to 2019, the expansion of its scale (e.g., in terms of GGR and international tourism expenditures as shown in Tables 1.2 and 1.3 in Chapter 1) in the ESEA as a whole and in various individual destinations has evidently produced the dynamics that further the development of this industry. To proceed, the existing industrial structures and organizations that evolved from the 1970s in the various destinations may have provided them with the basis of gaming that aligns with their social, political and economic environments. Nonetheless, in view of the foreseeable competition between the destinations and other factors that affect the demand for casino gaming, the structure and organization of this industry in the different ESEA countries/regions may still continue to change to ensure their sustained growth.

2.5.1 Existing structures and organization

Framed by the different public/political interests and legal systems at the outset, the structures of casino gaming in the ten existing legalized casino tourism destinations (see Table 2.2) share some similarities but are actually quite diverse in their practices.

Macao, Singapore and Malaysia

In Macao and Singapore, it is evident that an oligopolistic structure with large-scale and comprehensive IRs is adopted by the governments for the development of their casino tourism. In practice, since Macao and Singapore are both very small in their geographical area,[7] competition between the few large IRs in the market is inevitable. Nevertheless, while the number of casino gaming licenses issued by the Singapore government limits the number of casinos to be opened (i.e., two), each of the six casino gaming license holders in Macao is allowed to open more than one casino. Thus, a total of 30 casinos were operating on the main IR properties and individual hotels of the six licensed operators at the end of 2023. In contrast, a monopoly has been approved by the Malaysian government for the development of casino tourism in the country.

Table 2.2 Number of casino gaming licenses in various ESEA countries/regions (2023)

Country/region	*Number of casino gaming licenses**
South Korea	18
Macao	6[a]
Malaysia	1
The Philippines	50[b]
Vietnam	54[c]
Cambodia	85[d]
North Korea	2
Laos	3
Singapore	2
Japan	3[e]

Source: Author's online searches with cross-checking.

* Except for Macao, the number of casino gaming licenses indicates the number of legalized casinos in operation.

[a] In Macao, there is no formal restriction on the number of casino hotels or resorts that could be opened by a casino gaming license holder. At the end of 2023, 30 individual casino hotels and integrated casino resorts are operating in Macao.

[b] In 2017, 41 casinos in the Philippines were directly operated by the government-owned PAGCOR, and 9 were operated by privately owned firms.

[c] In 2019, there were eight casino licenses. In 2022, the number of casinos significantly increased mainly due to the government's approval of new licenses to many five-star hotels to introduce casino-type electronic gaming machines to foreign visitors.

[d] Includes 9 full casino licenses and 76 temporary casino licenses (https://kiripost.com/stories/cambodia-casino-cgmcc-gambling-licenses). Also, it is known that licensed land-based casinos are also simultaneously operating online casinos.

[e] At the end of 2023, one of the three licenses authorized in the gaming law was approved to an IR operator. It is expected that the other two licenses will be granted in 2025 or later.

Laos and North Korea

Although there are three casino hotels/resorts in Laos and two in North Korea, their industrial structure is more appropriately described as a regional monopoly instead of an oligopoly. The reason is that each casino hotel/resort is located far away from another in different regions in their countries and hence would not compete with each other directly in the local market. Besides, unlike other destinations, the locations of most casino hotels/resorts in these two countries are close to the borders of their neighboring countries (i.e., Thailand and China) where casino gaming is illegal.

South Korea

Although there are 18 casinos in South Korea, the industry is actually dominated by Kangwon Land, which is the only venue where local Koreans can participate in gaming activities. In terms of its GGR, Kangwon Land Casino accounted for over 50% of the total casino gaming turnovers in the country (50.6%) in 2019. In practice, Kangwon Land Casino can be said to be a monopoly in South Korea for domestic tourists, while the other 17 casino hotels/resorts are comparatively smaller in scale and competing in their respective tourism districts. Nevertheless, the opening of a new IR in 2024 will likely change the course of the frontiers of casino tourism in this country.

The Philippines, Vietnam and Cambodia

Unlike Macao, Singapore, Malaysia, Laos, North Korea and South Korea, a major difference found in the Philippines, Vietnam and Cambodia is that there are many approved casino hotels/resorts (50 in the Philippines, 54 in Vietnam and 85 in Cambodia as shown in Table 2.2). Nevertheless, the structure of the industries and the organization of the businesses are clearly different between these three casino tourism destinations. As mentioned earlier, most casino hotels/resorts in the Philippines were operated by the state-owned PAGCOR but it is in the process of privatization in the mid-2020s. On the other hand, although the reported number of casinos in Vietnam increased significantly in the beginning of the 2020s, most of them received approval to operate gaming machines only. As for Cambodia, the rapid expansion of its casino tourism industry in the 2010s was actually driven by foreign capital but the integrity of the business organizations is clearly in question.

Japan

Lastly, Japan is an up-and-coming casino destination, and it is anticipated that a maximum of three casino licenses would be granted by the government, with the first already granted to an IR complex in Osaka by 2023. The oligopolistic structure of casino tourism in Japan is more accurately interpreted as a kind of regional monopoly in three different destinations within the country. For example, while there will be only one IR in Osaka with a casino, the other two may be potentially developed in Tokyo, Fukuoka, Nagasaki, or Hokkaido, etc. In practice, however, it is evident that the potential business organization in Japan will be a different form from that in Macao and Singapore.

2.5.2 Probable changes

Considering the evidence mentioned earlier (including evidence discussed in Chapter 1), it is reasonable to suspect that the existing scale and industrial structure of casino tourism in the various destinations in the ESEA will most likely enter into a consolidation period in the 2020s although there might be expansion of some individual properties and introduction of new projects. Besides, despite that the various destinations report high annual growth rates in business turnovers in the post-COVID-19 pandemic period between 2022 and 2024 owing to the low base effect in 2021 and 2022, the growth rates in most destinations in the second half of the 2020s as compared to the 2010s are likely to slow down. Nevertheless, industrial organization and business routines will (and have to) experience some radical adjustments if sustained growth is to be a long-term strategy on the agenda of the ESEA governments and industry practitioners.

Actually, two distinctive factors have arisen in the 2010s which will definitely influence the future development of casino tourism in the ESEA. These factors include (1) the changes in the attitude of the Chinese government in regulating the cross-border gaming activities of its people and (2) changes in the gaming behaviors and preferences of the younger generation in the ESEA as a whole.

Changes in regulations in China

To a large extent, it is undeniable that the rapid expansion of casino tourism in the ESEA from 2000 to 2019 has been driven by the exorbitant spending of Chinese patrons in gambling activities. In 2020, a Chinese official from the International Cooperation Department of the Ministry of Public Security indicated that "about RMB1 trillion (*US*$145.5 billion) in funds flows out of China into gambling activities every year, underlining the issue as a 'threat to the country's economy and national security'" (Moura 2021). To restrict the cross-border gambling activities of the Mainland Chinese, the Chinese government has amended Article 303 of its Criminal Law to prohibit any form of activity conducted by anyone or entity (including foreign casinos) for "organizing or soliciting Chinese citizens" (IGA 2020) to gamble (in casinos) outside of Mainland China. To this, any guilty parties will be "jailed for between five and 10 years" (ibid.). Together with other policies, it is evident that the ESEA governments and IR operators need to sit back and reconsider the pace of their expansion and their industrial organization and business routines.

Changing gambling behaviors and preferences of the younger generation

Parallel to the rapid advancement in technologies from the popularization of the internet in the 1990s to the development of the digital economy in the twenty-first century, the gaming behaviors and preferences of the younger generation (i.e., those who were born in the 1980s and 1990s or Generation Y or the Millennials) also exhibit clear differences as compared to their parents. According to a commentary, an industry expert stated that the younger generation or "[millennials] grew up in entirely different technological surroundings and thus have entirely different gaming interests" (GGRAsia 2015). For example, they may not necessarily be interested in traditional gaming tables and gaming machines. Thus, continuous efforts made by the industries to redesign the layouts of their gaming floor and gaming products are necessary "to attract younger customers" (GGRAsia 2015) and hence sustain growth.

In addition, the changing lifestyle of the even younger Chinese cohort, that is, those born in the late 1990s or Gen Z or the post-millennials, who grew up in the digital economy, may actually accelerate the changes. Evidence shows that gambling is not necessarily their priority when they travel to casino tourism destinations like Las Vegas and Macao. However, if "they do gamble, they spend less money or want games very different from the conventional offer" (Meneses 2023, 62). Thus, changes in the gaming behaviors and preferences of the younger generation who are situated in the digital world will definitely push changes in the industrial organization and business routines of the casino tourism destinations, and compel operators to make some significant adjustments in their business products and facilities in the coming decades.

2.6 Summary

The facts, evidence and arguments presented in this chapter fairly explicate the SCPE framework as depicted in Figure 1.2 for developing the frontiers of casino tourism in the ESEA. In a nutshell, the geographical location, social and cultural settings, and historical progress of the societies in the ESEA differentiate the essence of their gambling activities with those conducted in the Western societies, and hence the process for the evolution of casino gaming as a formal business sector blended with tourism for the development of casino tourism (in contrast to the traditional and monotonic organization of gambling tours) as a modern industry.

Besides, due to the different political regimes and legal systems of the ESEA countries/regions, the advancements, industrial structure and organization of the modern development of their casino tourism also vary. Aside from internal factors, the rapid growth and development of casino tourism in the ESEA from 2000 to 2019 were largely due to the demand from the Mainland Chinese and supplemented by the spending of local patrons and those from the neighboring regions where casino gaming is illegal. In this process, the expansion of large-scale IRs with high-quality gaming and non-gaming facilities also undeniably facilitated the development of this industry.

After decades-long of development, the existing industrial structure and organization of casino tourism in the ESEA countries/regions are realistic scenarios that parallel their individual contextual settings. Nevertheless, ongoing and radical changes in the industrial structure and organization and business routines in the coming decade are inevitable, in consideration of the dynamics associated with the changing business environment, especially the changing political interests and policies of governments (including the inclinations of the local governments and political parties, and in particular the Chinese government policies for supervising the cross-border gaming activities of the Mainland Chinese), as well as the gaming preferences of the younger generation in the ESEA.

Notes

1 Except Japan, whose industrial revolution started in the early 1870s right after the country's Meiji Restoration in 1868 (see, e.g., Sumikawa 1999).
2 Details about the political regimes are consolidated by author from various printed and online materials.
3 Summarized with online information posted on the University of Melbourne library website "Southeast Asian Region Countries Law." Available at https://unimelb.libguides.com/c.php?g=930183&p=6721948. Accessed December 10, 2022.
4 The materials presented in this section frame the arguments that will be presented in the remainder of this book. Of course, these materials are far from sufficient in offering a complete pictures of the cultural attributes, social structures, political regimes and legal

systems in these countries. Nevertheless, readers who are interested in the related topics can easily use different sources to obtain more information.

5 See, for example, Yash and Anmol (2018) for the modern developments of the four little dragons: Hong Kong, Taiwan, South Korea and Singapore in the 1950s and the 1960s; Noland (1990) for informative and comprehensive discussions about the industrialization and development of the four little dragons and the ASEAN-4: Malaysia, Thailand, the Philippines and Indonesia, between the 1950s and 1980s.

6 According to a report by McKinsey & Company (2023), "Mainland China had the largest outbound travel market in the world." In 2019, the total spending by outbound tourists from Mainland China was USD255 billion. As compared to the total spending of USD649 billion in 2001 (Lim and Wang 2008, 452), the compound annual growth rate in these 18 years is 17.5%.

7 As of 2022, the total land area of Macao, Singapore, Malaysia and Japan is 33.3 km^2, 734.3 km^2, 330,241 km^2 and 377,974 km^2, respectively (data for Macao: https://www.dsec.gov.mo/getAttachment/2849a869–3c22–4c77-b0ea-b60735fb60b6/E_MN_PUB_2023_Y.aspx; Singapore: https://www.singstat.gov.sg/find-data/search-by-theme/society/environment/latest-data; Malaysia: https://www.ekonomi.gov.my/sites/default/files/2023–01/MEIF_2022.pdf; and Japan https://www.stat.go.jp/english/data/handbook/pdf/2022all.pdf).

References

Barkema, Harry G., Xiao-Ping Chen, Gerard George, Yadong Luo and Anne S. Tsui. 2015. "West Meets East: New Concepts and Theories." *Academy of Management Journal* 58 (2): 460–479.

Chien, Grace C. L. and Cathy H. C. Hsu. 2006. "Gambling and Chinese Culture." In *Casino Industry in Asia Pacific: Development, Operation, and Impact*. Edited by Cathy H. C. Hsu, 201–224. Binghamton, NY: The Haworth Press.

Eadington, Williams R. 1999. "The Economics of Casino Gaming." *Economic Perspective* 13 (3): 173–192.

Firzanah, Hannah. 2021. "How Genting Highlands Became Successful." Available at https://www.alvaproductions.com.my/post/how-genting-highlands-became-successful. Accessed January 22, 2023.

Franke, Paul. 2022. " 'Nobody Came to Monte Carlo to be Board': The Scripting of the Monte Carlo Pleasurescape 1880–1940." *Journal of Urban History* 48 (6): 1247–1260.

GGRAsia. 2015. "Millennials and Casinos: Bridging the Generation Gap." *GGRAsia* (September 28, 2015). Available at https://www.ggrasia.com/millennials-and-casinos-bridging-the-generation-gap/. Accessed November 18, 2020.

IGA. 2020. "Amendment to China's Criminal Law Targeting Foreign Casinos to Include 5 to 10 Year Jail Term." *Inside Asia Gaming* (October 22, 2020). Available at https://www.asgam.com/index.php/2020/10/22/amendment-to-chinas-criminal-law-targeting-foreign-casinos-to-include-5-to-10-year-jail-term/. Accessed November 24, 2020.

Kale, Sudhir H. and Mark T. Spence. 2009. "Casino Customers in Asia Versus Western Gaming Jurisdictions: Implications for Western Casino Operators." *Worldwide Hospitality and Tourism Themes* 1 (4): 320–331.

Kingma, Sytze F. 2008. "Dutch Casino Space or the Spatial Organization of Entertainment." *Culture and Organization* 14 (1): 31–48.

Lee, Choong-Ki. 2011. "History of the Casino Industry in South Korea." *Journal of Tourism Sciences* (in Korean) 35 (1): 451–464.

Lim, Christine and Ying Wang. 2008. "China's Post-1978 Experience in Outbound Tourism." *Mathematics and Computers in Simulation* 78 (2008): 450–458.

Lopez, Matthew Benjamin P. 2014. "PAGCOR and the History of Casino Gaming in the Philippines." Available at https://www.scribd.com/document/444340820/

PAGCOR-and-the-History-of-Casino-Gaming-in-the-Philippines. Accessed January 22, 2023.
McKinsey & Company. 2023. "Outlook for China Tourism 2023: Light at the End of the Tunnel." Available at https://www.mckinsey.com/industries/travel-logistics-and-infrastructure/our-insights/outlook-for-china-tourism-2023-light-at-the-end-of-the-tunnel. Accessed June 23, 2023.
Meneses, João Paulo (co-ordinated). 2023. "Special Report: Chinese Millennials." *Macau Business* (October 2023): 47–76.
Moura, Nelson. 2021. "Gaming Operators, Junkets, and Private Citizens Should Carefully Study New Mainland Penal Code Amendments – Report." *Macau News Agency* (November 30, 2021). Available at https://www.macaubusiness.com/gaming-operators-junkets-and-private-citizens-should-carefully-study-new-mainland-penal-code-amendments-report/. Accessed December 15, 2021.
Noland, Marcus. 1990. *Pacific Basin Developing Countries: Prospects for the Future.* Washington, DC: Institute for International Economics.
Paulès, Xavier. 2010. "Gambling in China Reconsidered: Fantan in South China During the Early Twentieth Century." *International Journal of Asian Studies* 7 (2): 179–200.
Price, John A. 1972. "Gambling in Traditional Asia." *Anthropologica* 14 (2): 157–180.
Richard, Brendan. 2018. "Las Vegas: Pas, Present and Future." *Journal of Tourism Futures* 4 (3): 182–192.
Schwartz, David G. 2003. *Suburban Xanadu: The Casino Resort on the Las Vegas Strip and beyond.* New York, NY: Routledge.
Sumikawa, Shunsuk. 1999. "The Meiji Restoration: Roots of Modern Japan." Available at https://www.lehigh.edu/~rfw1/courses/1999/spring/ir163/Papers/pdf/shs3.pdf. Accessed January 14, 2020.
Thompson, William N. 2015. Gambling in America: An Encyclopedia of History, *Issues, and Society*. Santa Barbara, CA: ABC-CLIO. LLC.
Weber, Max. 1951. *The Religion of China: Confucianism and Taoism*. New York, NY: The Free Press.
Wilkinson, Barry. 1996. "Culture, Institutions and Business in East Asia." *Organization Studies* 17 (3): 421–447.
Wong, Anthony Ipkin and Ricardo Chi Sen Siu. 2015. "Casino Tourism: How Does Entrepreneurship Change Macao?" In *Entrepreneurship in Small Island States and Territories*. Edited by Bodfrey Baldacchino, 218–234. New York, NY: Routledge.
Yash, Dangayach and Gupta Anmol. 2018. "Four Asian Dragons – Evolution and Their Growth." *International Journal of Advance Research and Development* 3 (1): 158–162.
Zhang, Juan. 2017. "Introduction: Integrated Mega-Casinos and Speculative Urbanism in Southeast Asia." *Pacific Affairs* 90 (4): 651–674.

Additional readings

Hahn, Elliott J. "An Overview of the Japanese Legal System." *Northwestern Journal of International Law &* Business 5 (3): 517–539.
Klebanow, Andrew. 2018. "Cambodia Rising." *Inside Asian Gaming* (October 4, 2018). Available at https://www.asgam.com/index.php/2018/10/04/cambodia-rising/. Accessed November 26, 2018.
Siu, Ricardo C. S. 2013. "The Economics of Asian Casino Gaming and Gambling." In *The Oxford Handbook of the Economics of Gambling*. Edited by Vaughan Williams Leighton and Donald S. Siegel, 55–86. New York, NY: Oxford University Press.

3 Social economics and political economy of casino tourism

The paths of development and business performances of any socially controversial industry like pharmaceuticals/drugs, tobacco, alcohol and gambling are not simply determined by individual demand and supply in the respective markets. Actually, the behavior of consumers and firms and their interactions (economic exchange) in these markets are directly and indirectly influenced by the attributes and structure of a society, as well as the political interests and regulations of the government. To this, social economics and the political economy provide a set of realistic principles to examine and elucidate the development frontiers of the practices of these industries, and their changes over time, which, in this context, include casino tourism.

Recall the SCPE forces discussed in Chapter 1 (see the lower part of the analytical framework in Figure 1.2). The interactions between the social and political forces and their influences on the modern development of casino tourism (i.e., $a_1, a_2,$ *and* $a_3,$ *and* $s_1, s_2,$ *and* s_3 in Figure 1.2) are explicated in this chapter from the standpoints of social economics and political economy. To begin with, the rudiments of these two areas of economic studies are discussed in the first section. In particular, the related principles, arguments and approaches in social economics and political economy which are essential for exploring the practices and development frontiers of casino tourism are underlined.

Based on these principles, the impacts of the social and political contexts on the practices and progress of casino tourism in the major destinations of the ESEA are examined in the second and third sections, respectively. These include the interactions between different interest groups in the social development process and their influences on individual and public choices for the practices of casino tourism. In addition, the role of politics (i.e., the political system adopted in a society and how a society is regulated by its government) in the structure/organization of the industry and the evolution of its business routines are also elaborated. Moreover, possible changes in the path of development of the industry as a consequence of the dynamic interactions between social and political interests over time are analyzed in the fourth section.

Takeaways from this chapter:

1. Social economics examines how individual behaviors are influenced by group behavior which is characterized by ethnicity, culture, social norms and so on and

DOI: 10.4324/9781003279884-3

so forth in a society. The insights that arise from the interplay between individual consumers and firms in a dynamic environment for societal and economic changes are also provided.

2. The political economy scrutinizes the impacts of politics on economic activities. To do so, the interplay between the political interests/inclinations of a government and its laws and regulations, and the behaviors of the market participants and their impacts on the progress of an economy, is mainly examined.
3. Although existing studies on the impacts of socio-economic factors on gaming choices are mainly presented for the Western societies, findings on the negative relationship between education level and gaming spending, and positive relationship between income level and gaming spending, etc. are valuable references for operators in the ESEA.
4. Geographic factors (mainly the location of IRs and distribution of the Chinese community) which have significant impacts on the business turnovers of casino-type gaming in Australia have valuable implications on the destinations in the ESEA.
5. The popularity of the pachinko game in Japan is specific to its culture and its lack of popularity in Macao has important implications for IR operators in their future business development plans for Japan.
6. Government decisions at the outset determine the industrial structure and organization of casino tourism, while the contents and enforcement of related laws and regulations underlie the practices and business routines of the market participants.
7. The principles/reasons for government interventions could be similar across countries/regions, and the nuances of the actions and actual impacts are influenced by the respective social and economic settings in different jurisdictions.
8. The interplay between the practices of casino tourism and the social, political and economic factors in a dynamic environment produces ongoing forces which alter the organization and business routines of the industry.

3.1 Principles of social economics and political economy

Compared to orthodox economics (i.e., mainstream or neoclassical economics) today which mainly centers on the efficiency of the free market and rational choices of consumers and firms in the markets, most of the studies in the literature presented by the classical economists in the eighteenth and nineteenth centuries largely explored the impacts of the social structure and politics in a society on the practices of economic activities and the material well-being of the people.[1]

In congruence with the growing complexities of human society and its interplay with the economy in the twentieth century, academic interest in exploring economics-related topics (including economic development) from various social perspectives was revitalized in the 1940s and moved forward in the 1970s.[2] Social economics greatly emphasizes the interrelationship between market and social activities within the contextual settings of a society. For example, social economists commonly investigate the impacts of social norms, social development, education or demographic distribution on the economic behavior of consumers and firms and the path of economic development; or conversely, the influences of economic development on inequality, pollution, or crime.

As a major and most influential entity in a society, the government and its role in the practices of an economy (which contributed to the origins of studies in political economy) have long been an essential area of study by economists.[3] Despite that mainstream economists embrace and emphasize a free market, all market activities are indeed practiced within a society in which they are subjected to the laws and regulations that are government formulated and implemented at the outset. Thus, "political economy begins with the political nature of decision making and is concerned with how politics will affect economic choices in a society" (Drazen 2000, 5). In fact, political economy itself is a subject that is examined individually although the relationship between politics and economic activities is a topic found in social economics. In considering the essence of social economics and political economy, their analytical frameworks and approaches provide use information to reveal the development frontiers of casino tourism.

3.1.1 Analytical framework of social economics

In the modern literature of social economics, the significance of social factors like the "culture, norms, and social structure" (Becker and Murphy 2000, 3) of a society on the economic behaviors of consumers and firms is a commonly investigated topic. For example, it is identified that "a teenager's propensity to take drugs and to smoke is very much affect by whether his peers do; . . . activities, behavior and consumption most subject to strong social pressures from peers and others are those that take place publicly" (ibid., 3–4). Although Becker and Murphy (2000) and other studies in general in the social economic literature do not include gambling and casino tourism as examples, the observed behavior and choices of consumers who partake in these activities are unarguably influenced by the set of social factors.

On the basis of a comprehensive review of the existing literature on social economics, Benhabib et al. (2011) neatly categorize the studies/approaches of social economics into "three subfields," which include "Social preferences, Social actions, and Peer and neighborhood effects" (ibid., xviii). In studies of consumer behavior, a key incorporated notion of "**social preferences**" is that individual choices in the markets are not simply regarded as natural and the same, but nurtured and diverse. "**Social actions**" emphasize elucidating how "agents interact 'socially,' and their actions at equilibrium are social in the sense that they are not mediated exclusively by markets" (ibid., xix). In addition, related studies explore how the collective actions of different groups exert impacts on the choices of each other in the markets.

In terms of "**peer and neighborhood effects**," impacts from social networks (like family members and friends from the living community, school, church etc.)

on social and market behaviors have been investigated. These include the social phenomena of criminal and addictive behaviors (ibid., xx). To a large extent, it is clear that while the focuses of these three subfields of social economics are not exactly the same, they share a common ground in their analytical frameworks. That is, when examining the behaviors of market participants, the impacts of social factors are taken into consideration as endogenous variables instead of exogenous variables, which is commonly assumed by the orthodox economic approach.

3.1.2 Essence of political economy

By referring to the various political economy doctrines and approaches, it is possible to observe that the different arguments posed by world-renowned thinkers/economists on the anticipated role of a government and the impacts of its policies on the economy are contextual and analytical, which are specific to their lived eras and societies, respectively.[4] For example, although the thoughts of Adam Smith (1723–1790) and John Stuart Mill (1806–1873) on the political economy are partially inspired by the Industrial Revolution between 1760 and 1840 in England and the social and economic processes right before and after this era, their lived periods and the differences in their stage of academic work, work experience and knowledge associated with different phases of the Industrial Revolution actually led to different focuses in their arguments on the role of governments in a free-market economy. In contrast, the standpoint of John Maynard Keynes (1883–1946), who argues that the government is responsible to apply its fiscal policy for ensuring full employment, is largely influenced by his experiences living in the Great Depression in the 1930s.

Actually, the contextual nature of political economy implies that despite its core priority and the endeavors of economists to *investigate the relationship and interplay between politics and economic activities*, the actual topics of interest and exploration greatly vary over time and across different societies and industries. Besides, the approaches used to address the different topics are also quite diverse between the different schools of economists (e.g., advocates of neoclassical economics versus Keynesian, Marxian and institutionalism). Alternatively, political economy has been vaguely defined, which gives it "the virtue of being all-inclusive" (Drazen 2000, 5) as long as a robust and recognized approach is used.

Following the rapid expansion in the different types of industries and economic activities in the second half of the twentieth century (especially the advancement of various service industries like banking and finance, information and technology, tourism and gaming), there has clearly been an increase in academic interest around political economy studies at the industry level (which is different from the society/economy level as per Smith, Mill, Keynes etc.).[5] Considering the nature of casino tourism to which the free market may not be the ideal structure nor approved by the general public and government in most (if not all) jurisdictions, the political economy would definitely comprise an essential framework to explore the interplay between politics and economic activities for the sustained growth of this industry and its interrelationship with the related sectors in the economy.

3.2 Socio-economic attributes and impacts of casino tourism

Subsequent to the modern development of casino tourism, an increasing number of discussions on the practices of this industry have been conducted from a socio-economic standpoint. In particular, different social groups have inquired about the socio-economic factors that influence the individual choices for casino gaming, which is also explored by the academics. As a matter of fact, individual choice of casino tourism could be fairly explained by the principles of social preference and the peer and neighborhood effects. Besides, while the changes in socio-economic factors may have different impacts on the demand for casino tourism, the marginal effects of these factors on the demand for traditional forms of casino games, especially physical table games or slot gaming machines, may diminish or even be negative as a consequence of social development. On the other hand, practices of casino tourism may also generate feedback effects which interact with the development of a society (i.e., feedback through e and a_4 in Figure 1.2 on the ongoing progress of the industry) and hence changing the path of development of this industry.

3.2.1 Significance of socio-economic factors

To understand the frontiers of the development of casino tourism, a set of socio-economic factors that influence/determine the demand for casino gaming need to be described. Although empirical studies of these factors in the destinations in the ESEA are yet to be available, related analyses have been recently performed in some of the Western countries like Australia and Germany to reveal the impacts of various socio-economic factors on the demand (measured by the expenditure of patrons) for casino gaming. Actually, the results derived from these studies provide valuable implications/references for future research work in the destinations in the ESEA. In these studies, factors like ethnicity, gender, age, family composition, income level, source of income, occupational status and education level are commonly used as the variables to reveal (explain) the differences in the spending of different groups of consumers with different attributes on casino gaming (see, e.g., Constanze and Luca (2019) for the case of Germany, and Layton and Worthington (1999) for the case of Australia). In general, it is found that the income of German tourists has a strong correlation with their spending on gambling, while "[a] higher education has a negative effect on the propensity to gamble" (Constanze and Luca 2019, i).

Layton and Worthington (1999) find that Asian households in Australia are more inclined to spend on casino-type games like blackjack and roulette instead of other gambling games like poker machines and lotteries of different forms in comparison to other ethnic groups, like those from Europe and Africa (Layton and Worthington 1999, 438). In addition, Layton and Worthington (ibid.) also find that "geographic variables are significantly associated with the probability of play" including playing different gambling games, among which casino games have the highest marginal effects[6] in Queensland, Tasmania and the Northern Territory. Although

Layton and Worthington (1999) do not explicitly state this, it is worth mentioning that Queensland and Tasmania are the closest in proximity to the three cities (Sydney, Melbourne and Brisbane) where over 70% of the Chinese Australians reside.

Although the casino tourism destinations investigated in Constanze and Luca (2019) and Layton and Worthington (1999) are not in the ESEA, the socio-economic factors, methods and findings are actually a valuable reference for those who are interested in conducting similar research work on destinations in the ESEA. Since socio-economic factors like age distribution, education level and income in most Asian countries are experiencing significant changes in the twenty-first century, their impacts on the sustained growth of this industry definitely have to be addressed proactively by casino tourism operators in the ESEA.

3.2.2 *Social preference, peer and neighborhood effects*

From a qualitative standpoint, the aforementioned statistical relationships found on the spending of various groups with different socio-economic attributes on gaming (including casino gaming) and the differences among them may indeed reflect the general influence of their respective social preferences (e.g., different ethnicities) on individual choices. Besides, it is observed that decisions to spend money on a casino trip (e.g., the kind of games chosen or the duration spent on gaming and

Figure 3.1 Pachinko parlor in Hokkaido, Japan

Source: Author's photos

non-gaming activities during the trip) may also be affected by the preference/experiences of family members, friends and colleagues to a certain extent.

To illustrate these effects, the preference of the Japanese for the pachinko game (a kind of slot machine with balls, which resembles a pinball machine, see Figure 3.1) is an interesting case. Although the pachinko industry in Japan is legally considered to be an entertainment industry rather than a gaming industry, the pachinko parlors are actually a kind of slot parlor in practice (similar to the slot machine section in a casino). According to Tanioka (2000, 3), "*[p]achinko* playing is, and has been, undeniably one of the social facts of Japan since its introduction after WWII." In the second half of the twentieth century, playing pachinko indeed reflected the Japanese culture and social norms (especially among the working classes) which "dominate their lives, determining even the pace of their daily activities" (ibid., 25).

Despite the popularity of pachinko in Japan, however, casinos in Macao that attempted to introduce this game in the 1980s and 1990s to Hong Kong and Mainland Chinese tourists failed dismally. A major reason for this lack of interest is the difference in the social preference between the Japanese and Chinese in this game. In general, industry practitioners in Macao commented that the selection/preference of games by individual patrons is largely influenced by their lived social environment and word of mouth from their peers. Thus, Hong Kong and Mainland Chinese visitors who travel to Macao mainly play table games, especially baccarat. Implications from this on the composition of games for future IRs in Japan have to be examined carefully by the operators, especially in formulating effective business strategies for a trade-off between domestic and international demand for casino tourism. Besides, changes in the preference of the Japanese for the pachinko game in the twenty-first century also have to be recognized by the IR operators in Japan (further discussion will be made in Chapter 8).

Other than the pachinko game in Japan, the popularity of baccarat in most casino tourism destinations in the ESEA, like Macao, South Korea, the Philippines and Singapore, indeed reflects the particular social preference and neighboring effects under the Chinese culture on the choice of gambling games. As LaTour et al. (2009, 492 – 493) find, the Chinese are "more interested in social games (like baccarat) than individual games (like slot machines)" to meet their preference "for social recognition" which "seems to stem from a communal background that highlights social interaction and their culture's views of luck, fate, and destiny." These findings are indeed in line with some of the other studies on the impacts of cultural attributes of the Chinese on their choice of gaming games.[7]

3.2.3 Impacts of casino tourism on social and economic development

As described earlier, the demand for casino tourism is influenced by a set of socio-economic factors which then contribute to the development of the industry. In a dynamic environment, however, the direction and magnitude of these influences may actually change, and the changes are mainly driven by two forces: (1) feedback effects from society on the practices of the industry and (2) social development. In terms of the practices of the industry, business strategies formulated and implemented by casino tourism operators to meet and cultivate demand also generate both positive and negative impacts on society and its economy (e.g., modernization of the society and income levels of the community, as well as the probable family issues and crime), which may change the existing impacts of the socio-economic factors (e.g., through the changes in the perspective and behaviors of the various social groups) on the development (and continuous expansion) of casino tourism.

On the other hand, throughout the process of social development which is commonly reflected by increasing education levels and enhanced quality of life, the perceptions of the communities with casino tourism may also change over time. These include their choice of different types of gaming games and their views about the progress of casino tourism in their own society. Actually, the ongoing interplay between the practices of casino tourism, feedback from society about the practices and social development constitute an essential component of the forces (another component is largely the political system, which will be explicated in the next section) that shape the path of development of this industry.

Some of the existing studies in the literature (see, e.g., Sims (2016) and (2017) for the case of Laos; Gunesch (2017) for the cases of Macao, Singapore and some of the other Southeast Asian regions; and Wu and Chen (2015) for the case of Macao) have identified both positive and negative impacts from the practices of casino gaming/tourism. Generally speaking, the local communities find that urban development is enhanced and there are more high-quality tourism facilities/attractions from the expansion of casino tourism. The successful experiences of Macao and Singapore have led to the observation that "governments and non-state actors across Southeast Asia have developed gambling establishments as a means of fast-tracking economic growth and stimulating national development" (Sims 2017, 675). Also, local residents like those in Macao and Singapore "affirmed a standard upsurge in social welfare" (Wu and Chen 2015, 291) owing to the increase in public revenue from gaming tax.

From a societal standpoint, however, the issues of most concern in the development of casino tourism are undoubtedly the inexorable likelihood of increasing the inclination of the local community to gamble, especially the working class and youths. In fact, social concerns about gambling addiction and related family problems, crime and other adverse social and ecological impacts have been generally raised by different social groups with reference to the negative evidence identified/reported (Sims 2016, 2017, 677 – 678; Wu and Chen 2015, 8–9). Besides, as Gunesch (2017, 9) argues, the practices of casino tourism have "to be more responsible, not just about their gaming decisions and influences on their social circles, but also regarding their sociocultural and ecological settings."

Thus, it is evident that despite the possible increases in the reported monetary revenue, social welfare and high-quality tourism/entertainment facilities for enjoyment by the local community, arguments about the downsides and unquantifiable adverse social impacts associated with the progress/expansion of casino tourism may actually limit the frontiers of the development of this industry. Certainly, some of these socio-economic forces would impact the practices of casino tourism though their influence on government policies and activities.

3.3 Political economy of casino tourism

In contrast to most traditional businesses like agricultural, manufacturing, retail and catering, casino tourism in any modern society is only legitimized after specific related laws of the governments are passed, which require a demanding evaluation process at the outset. Without this approval, a firm cannot incorporate casino gaming as part of its tourism and hospitality services for its clients to enjoy. In other words, legitimacy is a pre-condition for firms in this industry to operate their business. Besides, their operations are usually strictly monitored (or even periodically reviewed) by the government.

The legal authority of the government means that its power to legalize and draft gaming and related laws and regulations will shape the structure and organization of this industry. During this process, there is also the trade-off between the social and economic forces from the various interest groups (including social and religious groups and business parties) and the decisions made by the government which represent its political interests and inclination. As for the practices of this industry, the contents and characteristics of the laws and their implementation will indeed provide different incentives for the market participants as they establish their business routines since casino gaming is easily linked to other socially and politically controversial activities like gambling addiction, crime, usury, money laundering, corruption etc., including activities that operate in the gray areas of the law and legal system. To these, ongoing government interventions in the business routines of this industry are indeed the norm. There are many examples from the casino tourism destinations in the ESEA that support this argument.

3.3.1 Reasons and practices for government intervention

Early studies on the political economy of casino gaming and casino tourism in the United States of America (especially the cases of Las Vegas and Atlantic City) argue that the various forms of government intervention in the practices of the casino gaming industry strive to balance "two strong but conflicting forces" (Eadington 1984, 24). These refer to the actions taken by a government as a trade-off between economic benefits (e.g., gaming revenue and economic growth for the jurisdiction) and "potential significant risks to the social, political and moral fabric of the community and society" (ibid.) due to the practices of casino tourism.

On the basis of these primary considerations, Eadington (1998) also argues that governments in different states of the United States have to carefully examine a list of specific public policy questions to regulate casino tourism. These mainly include the objectives of the gaming laws and regulations, the desired structure and organization of the industry, the potential impacts from the accessibility of the gaming facilities by locals and tourists, the means for gaming promotion etc. (Eadington 1998, 54). Actually, these arguments of the benefits and costs/impacts associated with casino tourism and the primary policy considerations well explain the importance and areas of government intervention for the modern development of casino tourism, which are also applicable to other jurisdictions around the world.

For example, some of the IRs in the ESEA are only allowed to receive foreign patrons, while others may impose entrance fees for local visitors. Besides, some governments limit the number of issued gaming licenses in their laws, while others do not.

As discussed in Section 2.2.2 of Chapter 2 on the diversity of political regimes and legal systems across various countries/regions in the ESEA, one may suspect that even though there are similar reasons in general for government intervention in their respective jurisdictions (e.g., the interest in economic benefits like capital investment, gaming taxes and local employment, and concerns of social costs/impacts like gambling addiction, crime and corruption), the contents, scope and degree of the interventions differ among the different jurisdictions. Besides, the observed government activities do not simply reflect the political interests of this industry but are embodied by its considerations of the influences of the structure and stage of development of its society and economy, as well as the attitudes of various interest groups.

Taking into consideration gaming taxes as an example, there is a common consensus that the legalization and development of casino tourism would generate an attractive amount of tax revenue for government. Advocates argue that the development of most casino tourism destinations and the industry provides more revenue for government spending on local social welfare without the pressure to levy higher taxes in other areas (especially income tax) on the local residents (e.g., the evidence in Wu and Chen (2015) for Macao and Singapore) since mainly non-local tourists spend on gaming and non-gaming services. In practice, however, gaming and tourism tax rates vary across different jurisdictions. To a certain extent, these differences reflect the different roles and contributions that the governments expect from casino tourism for their economies and societies respectively.[8]

Besides, according to "Casinos as Special Zones: Speculative Development on Nation's Edge" (Zhang 2021), the decisions of various governments in Asia to approve casino tourism under the IR model from 2005 to 2019 exhibit public interest "to improve foreign direct investment, state tax revenue, local employment" (Zhang 2021, 73) through the modern development of this industry. During this process, however, ways in which the governments monitor their respective markets are different. For example, evidence shows that while "Singapore and Vietnam serve as examples of the state's strong regulatory presence in IR projects, states like Laos seem to have opted for a different approach by staying at arm's length from the day-to-day operation of these casino development zones" (ibid., 79).[9]

Apart from taxation and supervision, government interventions also include the establishment of requirements (through laws, regulations, administrative orders etc.) for the firms to carry out programs and initiatives like responsible gambling, social development, anti-money laundering, environmental protection etc. to minimize the possible adverse social impacts from their business operations instead of simply allowing them to profit from their ventures. Thus, it is evident that despite the potential demand and supply of casino tourism in the market, government interventions act as boundaries that limit this relationship. However, the interplay between the firms and government over time may change the boundaries.

3.3.2 Coverage and enforcement of laws on business routines

The contents (coverage and details) of gaming laws and regulations and their enforcement convey essential messages/incentives to firms which then use these to design and execute their business strategies. On the other hand, the industrial structure and organization of casino tourism in various jurisdictions in the ESEA are based on gaming laws that are constructed from their respective legal systems at the outset. Over time, interactions between the casino tourism operators and government during the enforcement of the related laws and regulations are an important influence on business practices and business routine changes in the industry.

For example, how casino gaming businesses are operated and the settings of the non-gaming businesses and their interplay with the gaming sector are quite diverse, although the gaming venues and some of the non-gaming facilities in the IRs across various jurisdictions like Macao, Singapore, the Philippines etc. share some similarities. In fact, the diversity is because of the different connections and ways of doing business between the gaming agents and the IR operators, and the distribution of business revenue between the gaming and non-gaming sectors of the IRs across the different jurisdictions (these differences and other phenomena that illustrate the diversity will be revealed in the following chapters in discussions of the individual jurisdictions).

Considering the aforementioned nature of casino tourism in practice, a negative correlation between the comprehensiveness of the contents of the related laws and regulations and the business turnovers in the industry is commonly suspected/observed by the general public in any given short period of time. This negative correlation is also applied to the actions taken by the government in the enforcement of the laws and regulations. Indeed, these two arguments are exemplified by comparing the historical evidence from various jurisdictions like Macao, Cambodia, Malaysia, Singapore and South Korea. In the long run, however, sustained growth and business expansion under an insufficiently and loosely regulated environment are always in question. A typical case for reference is the dramatic surge and slump of the casino gaming sector in Macao from 2000 to 2019 (even with the unexpected external shock from the COVID-19 pandemic between 2020 and 2022 taken into consideration; see, e.g., Siu (2023)).

In congruence with the rapid progress of casino tourism in the ESEA from 2000 to 2019, and the increasing competition in the region, it is evident that the governments of the various jurisdictions are striving to support the further development of their own industries on the one hand and strengthening their interventions on the other hand to ensure the competitiveness and hence sustained growth of the industry in the long run. Thus, a more dynamic role of the governments in the development of this industry is foreseen in the future. Nevertheless, finding ways to include and making trade-off between the interests and feedback of the various stakeholders to ensure the effectiveness of the interventions (in terms of adjusting the contents and enforcing the related laws and regulations) are definitely inevitable challenges faced by each and every government.

3.4 Dynamics of different interest groups

The discussions presented in Sections 3.2 and 3.3 show the influences of the various socio-economic and political factors on the development of casino tourism. In a dynamic environment, these factors and the practices of casino tourism are indeed mutually interplaying and producing continuous forces that evolve this industry. As shown, the legitimacy and industrial structure of casino tourism in a jurisdiction reflect government's decision (i.e., a public choice) in coordinating the interests of various social, political and economic groups at the outset. Subsequently, the practices and performance of the industry are, to a certain extent, determined by the contents of the gaming laws and related regulations and their enforcement by the government. In turn, feedback from society and response from the economy regarding the effects (both positive and negative) of the practices of this industry over time will explicitly exert pressure on the government to review and amend its laws and regulations. These will then lead to industry changes in terms of the existing scale and scope, business routines and even the path of development.

Besides, the interests of different groups from their respective standpoints on the impacts of the practices of casino tourism do not always align, and the differences are quite substantial in the real world. Thus, their ongoing tug of war with the practices of this industry will have direct and indirect, and positive and negative forces which then change the frontiers of the development of casino tourism. During the process, the impacts of the representative decisions of the government for society on the progress of this industry will never be without any contentions among the different groups. To this, there is no "optimal" industrial organization or business routine that would suit all of the casino tourism destinations nor any one destination on a continuous basis.

3.4.1 Appeal versus reality of the industry for advocates

Industry practitioners and governments who are in favor of the development and expansion of casino tourism usually emphasize the monetary benefits from this industry in their various forms (i.e., profit, income and tax) and financial resources for capital investment in facilities and infrastructure to enhance the tourism attractions and business efficiency, and increase in employment opportunities in the jurisdictions. Individual supporters in the concerned region or country also have self-motivated agendas. For example, increased government revenue from gaming taxes means increase in higher-quality social welfare programs, such as better education and medical services to the general public without the need to raise the income taxes of various local working groups (or real income tax payments can even be reduced). On the other hand, locals enjoy the high-quality entertainment and catering services of IRs at attractive prices (the casino gaming operations and spending of tourists subsidize the cost), while more job opportunities are provided to the community. Besides, local residents may also bask in the modern development of their city with a number of iconic IR buildings and related world-class facilities (e.g., the Marina Bay Sands and Resorts World Sentosa in Singapore,

Genting Highlands in Malaysia, Newport World Resorts (formerly Resorts World Manila) in the Philippines, Kangwon Land in South Korea, and a large number of landmark buildings on the Cotai Strip in Macao).

As for the advocates, although they recognize the negative impacts that extend from the practices of casino tourism like reported problems of addictive gambling and pollution of various forms, they argue that the gains from the operations of this industry are greater than the losses. From their position, for example, a government and the related operators could devote the needed resources to resolve some of the major social issues/concerns like addictive gambling, and the expenditures to contribute to the economy are only a small share of the benefits derived by the industry. In addition, they also argue that other social and political issues like crime, corruption and money laundering are alleviated by strengthening government supervisory power and comprehensiveness of the related laws and regulations.

3.4.2 Apprehensions and criticisms of opponents

The opponents (especially the different ethnic and religious groups) of casino tourism argue that the majority of the adverse impacts due to the practices and expansion of casino tourism on society can hardly be concretely determined (or are even immeasurable) and reflected in monetary terms. They believe that public expenditures on various treatment and preventive programs and corporate spending on social responsibility programs do not resolve the problems themselves, as they do not address the root of the problem. For example, unreported family problems and stress associated with individual gambling activities, underground economic and criminal activities abetted by the expansion of casino tourism are indeed widely found and cannot be easily remedied. Actually, some of the local newsletters occasionally report evidence of prostitution, drugs, loan sharks, kidnappings, money laundering, light and noise pollutions etc., which are all attributed to the expansion of casino tourism (e.g., Macao, the Philippines, Cambodia etc.). The various social groups argue in these reports that the integrity and morality of the community are undermined.

Other than the identified and suspected social problems related to the operation of land-based casinos, advancements in technology also lead to new issues associated with online casino gaming activities which are operated by the firms located in some of the casino tourism destinations as a promotional strategy to online players. Parallel to the rapid expansion of casino tourism in the ESEA in the 2010s, it is observed that an increasing number of firms and casino resorts located in some of the jurisdictions like the Philippines and Cambodia also provide online casino gaming services simultaneously. Nevertheless, the practices of online casino gaming in those jurisdictions may not easily avoid crime and social problems.

For example, the Philippine National Police-Anti-Kidnapping Group (PNP-ANG) reported that in 2021 and 2022, "[m]ost of the kidnapping cases in the country occurred in areas in Metro Manila where most Philippine Offshore Gaming Operators (POGOs) and casinos are located"[10] (Santos 2022). In practice, the perceived impacts of the general public on the development and expansion of casino tourism which create a particular environment for the popularity of online gaming and hence the observed crimes can hardly be denied convincingly although the causal relationship between reported kidnappings (and other crimes) and online gaming, and online gaming and casino tourism development in Metro Manila can hardly be quantified or verified.

By referring to the discussed evidence, counter forces can be realized which inhibit the continuous expansion of casino tourism that is generated not only within domestic boundaries but also from cross-border entities like neighboring communities and their governments (e.g., in China and Thailand). For example, considering the anti-corruption policies of the Chinese government (which also contain the spending of the mainland Chinese on gambling trips), the South Korean government decided to grant only one new casino license in 2016 although a total of two were announced in 2015 (Kim 2016). On the other hand, Macao is the most popular casino tourism destination for the Mainland Chinese but has been associated with increases in irregular gaming activities, so the Macao SAR government carried out a radical review and subsequently revised its gaming laws and regulations between 2019 and 2022 (see, e.g., Siu (2023)).

In addition, the Chinese government called for meetings in the 2010s with the governments of various casino tourism jurisdictions like North Korea and the Philippines to tighten the monitoring of their industries and prevent gambling-related problems from being "exported" from their jurisdictions to Mainland China. To a large extent, insufficient government oversight in some of the casino tourism destinations has indeed created an environment with questionable (or even illegal) casino gaming businesses which are controlled by organized criminal groups. Actually, related issues in some of the jurisdictions in Southeast Asia have caught the attention of the international community. Thus, the United Nations Office on Drugs and Crime (UNODC) also held a meeting (with representatives from non-casino tourism jurisdictions like China) in Thailand in 2023 (UNODC 2023) to draw the attention of the related nations to the issues.

3.5 Summary

Given the diverse social, economic and political structures and differences in the stages of the development among the different countries/regions in the ESEA (as elaborated in Chapter 2), the discussions and evidence presented in this chapter further show that the influences from a similar set of factors/variables (e.g., ethnicity, education, income) on the demand for casino tourism vary across different jurisdictions and over time in any of the individual jurisdiction. From the supply side, their practices are quite different, although the industrial structure and organization of this industry are determined by the legal decisions of governments at the outset. Other than market conditions, it is evident that differences in the contents and enforcement of laws and regulations produce different incentives to the market participants when they carry out their businesses and design their business routines.

As for the nature of casino tourism, it is also argued that feedback (which could be positive, negative or neutral) from the various interest groups on the practices of this industry and their interplay will produce continuous forces that evolve the business scale, scope and routines, and path of development. In this interacting process, opponents call for restrictions or even limiting casino gaming in terms of its scale and scope in the industry due to the adverse social impacts (on top of the monetary costs) that are identified with casino gaming activities, as well as irregular and illegal business activities of other third parties in the casino tourism destination while advocates usually tout the economic benefits from the practices of casino tourism.

In view of the aforementioned facts, it is obvious that the social, economic and political perceptions of casino tourism in the different jurisdictions in the ESEA greatly differ throughout their respective development processes. Taking into consideration the structure and realities associated with their own societies and the different degrees of influence and historical influence from the Western world, the interplay among the various stakeholders in these jurisdictions will definitely generate ongoing forces that impact the progress of the organizations and business routines in the industry over time. In the following chapters, the major topics emphasized in Chapters 1 to 3 will be examined and explicated in detail with reference to the development of casino tourism in the various major and emerging jurisdictions.

Notes

1 The publication of *The Wealth of Nations* by Adam Smith (the father of modern economics) in 1776 marks the beginning of classical economics. Although the arguments of a free market with limited government interventions in economic decisions/activities as Smith explicates in his treatise (Books I and II) are common knowledge, he also examines the impacts from the structure of and changes in the social environment and politics in a nation on the progress of the economy in detail (Books III, IV and V). Subsequently, the *Principles of Political Economy* published by John Stuart Mill in 1848 is another landmark classical literature which scrutinizes the essence and impacts of social structure and politics on a free-market economy.

2 A milestone in formal and systematic studies in social economics is the establishment of the Association for Social Economics in 1941 (https://socialeconomics.org/), and its ongoing publications of the journal *Review of Social Economy* since 1942 and then the *Forum for Social Economics* since 1971. Besides, another leading journal, the *International Journal of Social Economics*, in this area of study was introduced in 1974 (see, e.g., Thanawala 1985).

3 Scholarly interest in systematically examining the roles and impacts of government policies on production and exchanges within a country and trade between different countries has been found since the sixteenth century under the economic thoughts of mercantilism (arguments from this school of economic thoughts can be easily found with online searches; see, for example: https://www.britannica.com/money/topic/mercantilism).

4 This observation is made after reviewing the comprehensive discussions of the different theories of political economy which range from the classical approach to neoclassical, Keynesian and contemporary libertarian approaches (Caporaso and Levine 1992).

5 For example, the first publication of *A Companion to Tourism* (edited by Lew et al. (2004)) provides a comprehensive collection of contemporary studies in tourism and includes the article "Toward a political economy of tourism" (Williams 2004).

6 This means that if casinos are located in these areas, their revenue is higher as the patrons spend more money on gambling.

7 Owing to the changes in age, education level and other social attributes of Chinese outbound tourists, their choice of gaming games has also changed in the 2020s and onward (related discussions will be presented in Chapter 9). Thus, it is worthwhile highlighting that even though baccarat may still be the most popular and highest turnover game in the ESEA in the 2020s (or even the 2030s), its growth would likely diminish over time and, hence, IR operators need to make proactive business strategies that change the composition of gaming devices on their properties.
8 Differences in the existing tax system between different countries/regions also explain the different practices of their gaming taxes.
9 These differences will be further explored in the following chapters for the individual casino tourism destinations.
10 POGOs are the online offshore gambling firms registered and located in the Philippines and provide online casino-type gambling services to overseas players.

References

Becker, Gary S. and Kevin M. Murphy. 2000. *Social Economics: Market Behavior in a Social Environment*. Cambridge, MA: The Belknap Press of Harvard University Press.

Benhabib, Jess, Alberto Bisin and Matthew O. Jackson. 2011. "Social Economics: A Brief Introduction to the Handbook." In *Handbook of Social Economics*. Edited by Jess Benhabib, Alberto Bisin and Matthew O. Jackson, xvii–xxi. San Diego, CA: North-Holland.

Caporaso, James A. and David P. Levine. 1992. *Theories of Political Economy*. New York, NY: Cambridge University Press.

Constanze, Giebeler and Rebeggiani Luca. 2019. "Who Loves to Gamble? Social-Economic Factors Determining Gambling Behaviour in Germany." Available at https://mpra.ub.uni-muenchen.de/94735/1/MPRA_paper_94735.pdf. Accessed April 1, 2023.

Drazen, Allan. 2000. "What Is Political Economy?" In *Political Economy in Macroeconomics*. Edited by Allan Drazen, 3–19. Princeton, NJ: Princeton University Press. Available at https://assets.press.princeton.edu/chapters/s6819.pdf. Accessed January 29, 2023.

Eadington, William R. 1984. "The Casino Gaming Industry: A Study of Political Economy." *The Annals of the American Academy of Political and Social Science* 474 (11): 23–35.

Eadington, William R. 1998. "Contributions of Casino-Style Gambling to Local Economies." *The Annals of the American Academy of Political and Social Science* 556 (1): 53–65.

Gunesch, Konrad. 2017. "Socio-Economic, Cultural and Ecological Implications of Gaming and Tourism in Asia: The Pacific Region between Tradition and Modernity Regarding Comprehensively Entertaining, Family-Oriented and Culturally Preserving Heritage Travel." Available at https://doi.org/10.19275/RSEPCONFERENCES103

Kim, Eun-Jung. 2016. "S. Korea Gives Casino License to Mohegan-KCC Consortium." (February 26, 2016). Available at https://en.yna.co.kr/view/AEN20160226007551320. Accessed February 20, 2022.

LaTour, Kathryn A., Franck Sarrazit, Rom Hendler and Michael S. LaTour. 2009. "Cracking the Cultural Code of Gambling." *Cornell Hospitality Quarterly* 50 (4): 475–497.

Layton, Allan and Andrew Worthington. 1999. "The Impact of Social-Economic Factors on Gambling Expenditure." *International Journal of Social Economics* 26 (1/2/3): 430–440.

Santos, Jel. 2023. "Most of Kidnappings Are in NCR Areas with POGO, Casino Establishments." *Manila Bulletin* (September 25, 2022). Available at https://mb.com.ph/2022/09/24/most-of-kidnappings-are-in-ncr-areas-with-pogo-casino-establishments-pnp/. Accessed February 12, 2023.

Sims, Kearrin. 2016. "Laos Rolls the Dice on Casino Tourism." Available at https://www.westernsydney.edu.au/ics/news/blog/laos_rolls_the_dice_on_casino_tourism. Accessed February 11, 2023.

Sims, Kearrin. 2017. "Gambling on the Future: Casino Enclaves, Development, and Poverty Alleviation in Laos." *Pacific Affairs* 90 (4): 675–679.

Siu, Ricardo C. S. 2023. "Back to the Future: Constructing Macao as a World-Class Casino Tourism Destination under New Gaming Laws." Gaming Law Review: Economics, *Regulation, Compliance, and Policy* 27 (7): 326–342.

Tanioka, Ichiro. 2000. *Pachinko and the Japanese Society*. Osaka, Japan: Institute of Amusement Industries, Osaka University of Commerce. (English edition by: Osawa, Minae, Madeleine Vedel and Okamoto Miki).

Thanawala, Kishor. 1985. "What Is Social Economics?" *International Journal of Social Economics* 12 (6/7): 34–40.

UNODC Regional Office for Southeast Asia and the Pacific. 2023. "Southeast Asian Officials and China Meet with UNODC to Discuss Organized Crime Run Casino and Scam Operations." (June 28, 2023). Available at https://www.unodc.org/roseap/2023/06/organized-crime-casino-scam-operations-southeast-asia/story.html. Accessed July 30, 2023.

Williams, Allan M. 2004. "Toward a Political Economy of Tourism." In *A Companion to Tourism*. Edited by Alan A. Lew, C. Michael Hall and Allan M. Wiliams, 61–73. Malden, MA: Blackwell Publishing Ltd.

Wu, Shou-Tsung and Yeong-Shyang Chen. 2015. "The Social, Economic, and Environmental Impacts of Casino Gambling on the Residents of Macau and Singapore." *Tourism Management* 48 (2015): 285–298.

Zhang, Juan. 2021. "Casinos as Special Zones: Speculative Development on Nation's Edge." In *Development Zones in Asian Borderlands*. Edited by Mona Chettri and Michael Eilenberg, 73–94. Amsterdam, Netherlands: University of Amsterdam Press.

Additional readings

Alipour, Habib and Edwin M. Vughaingmeh. 2010. "Casino Gambling Tourism – A View from Community: The Case of North Cyprus (TRNC)." *Journal of Rural and Community Development* 5 (3): 175–202.

Aronovitz, Cory. 2002. "The Regulation of Commercial Gaming." *Chapman Law Review* 5 (2002): 181–208.

Kamat, Kustubh, Miriam Scaglione, Subhash Kizhakanveatil Bhaskaran Pillai and Ruey Feng Chen. 2016. "The Socio, Economic, and Environmental Impacts of Casino Tourism in Goa." *Revista Latino-Americana de Turismologia* 2 (1): 44–54.

Mosedale, Jan. 2014. "Political Economy of Tourism: Regulation Theory, Institutions and Governance Networks." In *The Wiley Blackwel Companion to Tourism*. Edited by Alan A. Lew, C. Michael Hall and Allan M. Williams, 55–65. West Sussex, UK: John Wiley & Sons, Ltd.

Tan, Danielle. 2015. "The 'Casino Strategy' in Laos and Cambodia: A Risky Bet on the Future." *GIS Asie* (July 1, 2015). Available at http://www.reseau-asie.com/article-en/months-articles-archive/reseau-asie-s-editorial/casino-strategy-laos-cambodia-risky-bet-on-future-danielle-tan/. Accessed October 23, 2021.

4 Reality or illusion? Specific lessons from Macao

The evolution of Macao in its transformation into a world-renowned casino tourism destination was a long process that started from the mid-nineteenth century when casino-type gambling games like lotteries were first approved by the Macao-Portuguese government in 1847 and then Fantan in 1849 (Godinho 2013). Following more than a century long of casino-type gaming and some simple/monotonic hospitality and entertainment options, casino tourism in Macao gained momentum with the promulgation of the first gaming law in 1961 and then explicitly formalized with the opening of a landmark casino hotel, the Hotel Lisboa, in 1970. Nevertheless, despite being dubbed as the "Oriental Monte Carlo," Macao was largely a casino gaming city in which tourism was an afterthought, which relied on (or was associated with) casino gaming rather than being its equal. Even worse, adverse social impacts that were increasing in severity from the practices of casino gaming under a lagging and loosely regulated system actually led to social chaos in the second half of the 1990s (Siu 2006).

Upon the establishment of the Macao Special Administrative Region (MSAR) of China under its "one country, two systems" policy when the Macao-Portuguese government formally returned Macao back to China at the end of 1999, the decision of the MSAR government to replace the monopoly structure of the industry with an oligopoly in 2002 unarguably opened a new chapter which propelled the development of Macao into a casino tourism destination. By 2006, Macao surpassed Las Vegas in reported GGR and became the world's largest casino tourism destination. On the other hand, an unprecedented upsurge in the macroeconomic performance through considerable amounts of capital investments by IR operators, rapid rise in public investment and social welfare spending by the government with gaming tax revenue, a healthy employment rate and so on and so forth caused excitement in many of the existing and potential casino tourism destinations in the ESEA.

Nevertheless, the dramatic performance of the industry was due to a large share of unorthodox and irregular business practices conducted by a particular group of highly independent gaming agents (i.e., the third-party operated business segment). In view of the changing path of development of the city as a whole and the need to ensure the sustained growth of this key industry for the economy, the MSAR government conducted radical revisions and promogulated revised and new gaming laws and regulations in 2022 to formally remove the persistent and questionable

DOI: 10.4324/9781003279884-4

business activities performed by independent gaming agents in 2023. Actually, these changes unarguably constitute as the new grounds for Macao to advance as a true world-class casino tourism destination in the long run (Siu 2023).

In this chapter, the particular contextual settings in Macao and the SCPE forces which brought about the expansion of casino gaming and its unique features for the modern development of Macao from a casino city to a casino tourism destination are explicated in the first section. Then, the unique factors that have contributed to the enviable business performance in the first two decades of the twenty-first century are revealed in the second section. Despite the exceptional performance of this industry, the adverse impacts from the practices of the third-party operated business segment on the sustained growth of the industry and the accumulation of counterforces for changes to be made are elaborated in the third section. Then, the essence and consequences of government direct interventions and revamping of the existing organization and business routines of the industry through the promulgation and enforcement of the revised and new gaming laws and regulations are discussed in the fourth section. Finally, it is argued that the progress of casino tourism in Macao actually provides a valuable lesson for the development of the frontiers of this industry, specifically in the ESEA.

Takeaways from this chapter:

1. The development of casino gaming and casino tourism in Macao balanced law and custom in the mid-nineteenth century which was largely influenced by the particular contextual settings of the city and their changes over time.
2. Evidence presented from the case of Macao confirms that the contents of a legal system and enforcement of the related laws and regulations are negatively correlated with the short-term business turnovers of casino tourism, especially casino gaming.
3. The "Chinese demand," a dynamic and controversial supply-side factor (mainly the third-party operated gaming businesses), and a lagging and loosely regulated system all contributed to the enviable and dramatic business performance of casino tourism in Macao from 2005 to 2019.
4. The Macao model is largely defined as highly independent third-party operated casino gaming businesses operating within the properties (VIP gambling rooms) of casino concessionaires which then informally share the GGR with the concessionaires.
5. The downsides of the Macao model are the unorthodox and irregular/illegal business activities undertaken by a particular group of gaming agents in a loosely regulated (or lawless) area to reap the economic benefits associated with casino gaming.
6. The transformation of the Macao model over time has created new challenges for society and the regulators.
7. The third-party-operated business segment was in principle terminated in 2023.
8. The revamping of the organization and business routines of the industry in 2023 through radical revisions of the related laws and regulations would unarguably point to a new and sustainable path for the long-term development of this industry.

4.1 From casino gaming to casino tourism destination

The progress of Macao from a gambling hub to a casino gaming city between the 1850s and 1950s, followed by its modern development from a prototype of a casino tourism city in the last four decades of the twentieth century to a casino tourism destination in the twenty-first century, has been largely contingent on the unique contextual settings of the society and changes in the SCPE forces over time (see, e.g., some of the major facts recapped in Godinho (2012)). Alternatively speaking, since the casino gaming industry is a key income source for different interest groups and the economy, changes in its path of development historically reflect certain major societal and political interest changes associated with the social development process from one stage to the other (i.e., largely reflect the changes in a_1, a_2, a_3 and s_1, s_2 as indicated in Figure 1.2 in Chapter 1).

4.1.1 Unique historical and contextual settings for casino gaming

Historically, the Portuguese developed Macao into a port for trade and transshipment between Western Europe and East Asia (especially China and Japan) starting in the mid-sixteenth century until the mid-nineteenth century (Wills and Dyke 2000).[1] However, this economic function/role of Macao came to an end when the British government formally opened a port in a neighboring city, Hong Kong, in 1842. As a tiny city (total land area in the last quarter of the nineteenth century was less than 10 km^2) with little real economic means aside from being a free port and a Chinese-based society, the Macao-Portuguese government and related interest groups saw gaming as a realistic/feasible and immediate opportunity to retain their economic benefits in Macao. Thus, the Macao-Portuguese government legalized gaming in Macao in 1847.

Nevertheless, although gaming businesses were approved through government-granted individual contracts in the early days, and then through an "administrative law concept of *concession*" under which "a temporary contract between the Government and a private entity that has been selected by means of a public tender" (Godinho 2014, 2) in various periods of time in the twentieth century, no gaming laws and regulations were formulated until 1961 (Eadington and Siu 2007, 4–5). This shows that the gaming business in Macao was indeed evolving and practicing in an environment that was loosely regulated. As a result, the behavior of the participants in the market was largely shaped by Chinese customs (i.e., the Chinese culture and their ceremonial ways of doing business) and their interplay with the Macao-Portuguese government based on its economic and political interests and legal system.

Other than its particular economic force (reality) and political and legal environments, the progress of casino gaming in Macao is also clearly framed by its social structure and development over time, and noteworthy are its unique geographical location and connections with the Chinese community in mainland China and Hong Kong. Geographically, Macao is part of Zhuhai city in the Guangdong province. Even before the Chinese government resumed sovereignty over Macao at the end of 1999, Macao already had close ties with different cities in the Guangdong province. Thus, the geographical location and social connections of Macao provided the city with a unique opportunity (advantage) to meet the demand for casino gaming as long as there was no competition from the neighboring regions, where, indeed, casino gaming is banned.

However, it must be noted that the attributes of the demand by the Chinese for casino gaming are primarily and specifically based on their cultural (including religion) practices in an underdeveloped social environment throughout its history, which are quite different from the Western societies. For example, although the rulers of various dynasties in ancient China might have promulgated codes of conduct to prohibit gambling of the masses, the "enforcement against 'gaming' was usually weak" (Wu and Lau 2015, 211). As a high form of culture (ibid., 212) which was practiced in a feudalist society, gambling was hardly "regulated" in Chinese history. Besides, owing to wars and social disturbances between the mid-nineteenth and mid-twentieth centuries, gambling of various forms became a common social phenomenon among the Chinese to speculate (irrationally) for a better life. Thus, as a land that was free but loosely/poorly regulated under the Portuguese administration, new business opportunities were made available to various groups to reap the economic benefits by "relying heavily on such sources of revenue as the pacapio lottery, the game of fantan, the sale of opium, the coolie trade, and later the vaeseng lottery" (Godinho 2013, 110).[2]

Evolving from the aforementioned contextual settings, it is evident that the development of casino gaming and casino tourism in Macao was actually practicing between law and custom. To a large extent, the modern development process and changes in the business organization of casino tourism in the last four decades of the twentieth century and the first two decades of the twenty-first century were highly influenced by these settings (Eadington and Siu 2007). Although the specific performance of the industry might vary in different eras, the differences were indeed embodied by the interactions between the different interest groups (including the government, firms and their associates) in congruence with the changes in the SCPE forces.

4.1.2 Attributes of modern development of casino tourism

Although the different casino gaming operators offered some cursory forms of accommodation and catering facilities in the first half of the twentieth century, the modern development of casino tourism actually unfolded in the 1960s. This is embodied in the success of the well-known Sociedade de Turismo e Diversões de Macau (STDM) in obtaining its monopoly position in 1962 following the promulgation of the first gaming law by the Macao-Portuguese government in 1961 (hereinafter "Gaming Law 1961"). In fact, the promulgation of this law represented an advancement in the legal system which not only improved the legal framework for the modern development of casino gaming but also explicitly spelled out the interest of the government that a successive gaming monopoly in 1962 was supposed to simultaneously contribute to the tourism development of the city (Eadington and

Siu 2007, 11–13). On the other hand, the success of the STDM was actually realized by the entrepreneurship of its founders (Wong and Siu 2015).

For example, to facilitate the arrival of visitors from the neighboring regions, especially visitors from Hong Kong at the time, the STDM established its own shipping company (the Shun Tak Shipping Company) in 1961. Besides, right after obtaining their gaming license, the STDM started to offer more entertainment facilities, which included the opening of a new casino (the Palace Casino) in a floating boat (the Macau Palace) in 1962 which was moored in the outer harbor of the Macao Peninsula (Pascoal 1995, 32). In fact, the operations of the Macau Palace also brought about some simple tourism-related business opportunities to the nearby streets. Then, investment in the construction and opening of a Western-style casino hotel in 1970, the Hotel Lisboa, was unarguably a milestone for the modern development of casino tourism in Macao. Through the opening of the Hotel Lisboa, the STDM also obtained government approval for the first time to install slot machines and not only table games in its casinos in 1972.

Together with investments in other new properties, the STDM experienced considerable business turnover from its casino gaming business. For example, it was estimated that between 1962 and 2000, the STDM obtained approximately USD25 billion (Liu 2002, 318–324; USD1 = Macao currency, MOP8) in total GGR (in current prices). With the financial power and ambition to preserve its monopoly position, the STDM also contributed to the urban development of Macao[3] directly and indirectly (e.g., through its Shun Tak Group). Thus, Macao unarguably increased its hospitality capacity and attractions (mainly through investment in various hotels, catering and entertainment facilities, as well as infrastructure) to tourists over time. Indeed, the total number of visitor arrivals increased by almost sixfold from around 703,000 in 1960 to 4,039,000 in 1980 (Liu 2002, 308), which was further increased to 9,162,200 in 2000.

Despite the positive contributions of the STDM to the modern development of casino tourism in Macao and of Macao itself, its business performance was actually underpinned by an innovative but controversial form of third-party-operated VIP-gambling-room business model (the "Macao model," which will be discussed in Section 4.3) that was introduced in the 1980s. After two decades of expansion, over 65% of the reported GGR in the industry was indeed derived by this particular "VIP" sector in 2001 (which is largely reflected in the periodical statistics for "VIP Baccarat" published by the Macao government). Through this particular business segment, Macao's reported GGR was around half of that of Las Vegas (the world's most renowned casino tourism destination at the time), although Macao only had a land area of less than one-tenth of Las Vegas in 2001. Nevertheless, irregular (or even illegal) business activities undertaken by third parties (which also involved triad groups) in a loosely/poorly regulated environment in pursuit of larger wagers from patrons ultimately led to almost uncontrollable social chaos in the second half of the 1990s (Brady 1998; Siu 2006; see discussions presented in Section 4.2.2), which clearly produced the counterforces for changes to the existing path of development of this industry.

4.1.3 Toward a casino tourism destination

Although the practices of the STDM unarguably led to the modern development of casino tourism in Macao, its third-party-operated business model most notably received negative feedback from society. To address the adverse impacts on the sustained growth and further development of this industry, the newly established MSAR government made the decision to replace the monopoly structure with an oligopoly by inviting international tenders for three new gaming licenses starting from 2002. To this, a license was granted to the world-renowned Wynn Resort and another to a joint venture between a Hong Kong-based entertainment company (Galaxy Entertainment) and another world-renowned mega casino resort (Venetian of the Las Vegas Sands) – Galaxy JV, while the third was approved to the newly formed Sociedade de Jogos de Macau S.A. (SJM), a wholly owned subsidiary of the STDM.

Shortly after the approval of these three new licenses, the MSAR government further exercised its administrative power to permit a major shareholder of each of the three corporations to develop their casino gaming businesses independently. Therefore, a total of six casino gaming/resort operators (Galaxy-Venetian, SJM-MGM and Wynn-Melco) launched their business in Macao in the mid-2000s. Despite the legal and social controversies of this change from three licenses to six individual operators, "a new arena for the world-class gaming entrepreneurs to effect their innovative ideas for the integrated development of casino gaming with tourism and other related service sectors" (Wong and Siu 2012, 222) was actually established.

Unlike the previous stages of development, large-scale capital investment in various comprehensive mega casino resort properties (in contrast to the one Hotel Lisboa and some small individual properties operating in the STDM era) unarguably transformed Macao from a simple casino tourism city into a destination where visitors could spend money on a large variety of non-gaming activities. In this new era, the development of large and comprehensive mega casino resorts (or IRs) actually made them the "destination" to visit for tourists regardless of whether their primary purpose for traveling is for gaming. Accordingly, the development of Macao as a casino tourism destination in 2000 to 2019 experienced unprecedented growth. However, concurrent changes in the irregular third-party-operated business activities under a lagging and loosely regulated regulatory system (Siu 2023) to pursue monetary benefits from casino gaming continued to challenge the sustainability of the path of development of this industry.

4.2 Factors that have led to enviable performance in the new era

To a large extent, the progress of casino gaming in Macao and its casino tourism prototype in the twentieth century were largely represented by a steady growth in monetary value with some but not wide-ranging development in the real capacity of the industry (e.g., certain capital investments made by the early gaming operators in the hospitality fixtures and later by the STDM in Hotel Lisboa and some small-scale modern tourism facilities). In contrast, development in 2000 to 2019 was not simply exhibited in the dramatic and unprecedented growth in the industry GGR (as depicted in Figure 4.1) but also a large amount of capital investment (over *USD* 100 billion between 2002 and 2019) spent concurrently by the casino resort operators on a wide range of comprehensive tourism attractions and world-class hospitality and entertainment facilities. Generally speaking, the factors/forces that led to the enviable performance of this industry in Macao in this new era are the same as those historically (i.e., the SCPE forces), but the practices and specific representations of those forces are quite different.

First, the political interests of the former Macao-Portuguese government and the newly formed MSAR government of China in the long-term role of this industry for the economy and society were clearly different while the change in the industrial structure from a monopoly to an oligopoly in 2002 was clearly an administrative rather than a market decision. Second, following the establishment of the MSAR, its social connections with mainland China accelerated and with the total support of the Chinese government for the continuous development of its casino tourism, demand from more Chinese income groups for casino tourism in Macao was found and increased dramatically (reflected by the increasing number of visitor arrivals from mainland China, as depicted in Figure 4.2).

Given the demand, the supply-side factors were also particular. On top of the trade-mark effects (e.g., the market excitement that world-class casino tourism services would be made available by the Las Vegas mega casino resorts in Macao for patrons in Asia) and expanding hospitality capacity, reorganization and practices of the third-party-operated business segment under a lagging and loosely-regulated

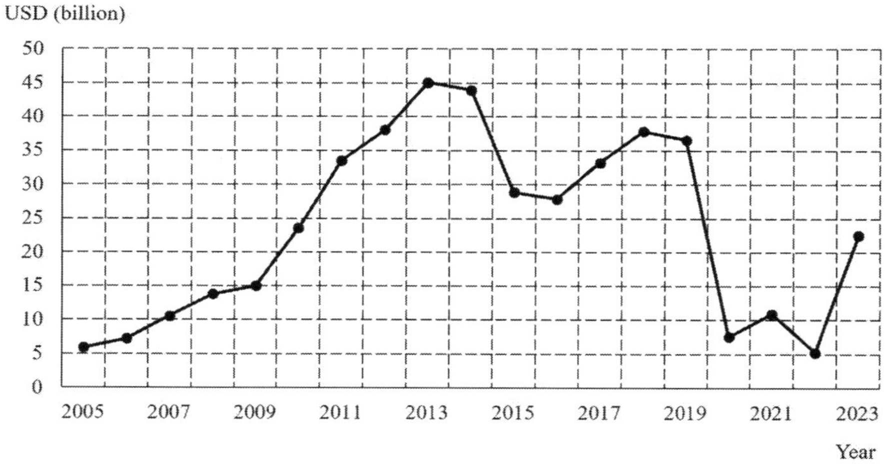

Figure 4.1 Changes in Macao's gross gaming revenue (2005–2023)

Source: Author (data derived from the website of Gaming Inspection and Coordination Bureau (DICJ) of MSAR, https://www.dicj.gov.mo/web/en/information/index.html)

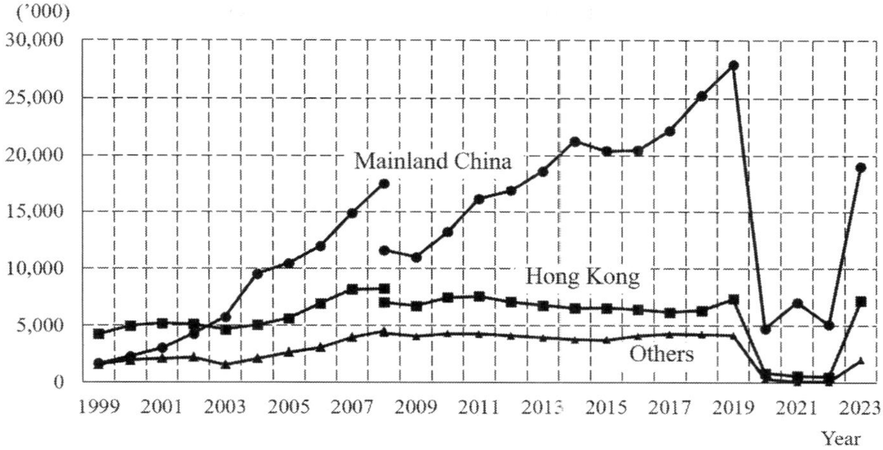

Figure 4.2 Changes in the number of visitor arrivals to Macao (1999–2023)[4]

Source: Author (data derived from the website of the Government of Macao Special Administrative Region
Statistics and Census Service (DSEC): https://www.dsec.gov.mo/en-US/Statistic?id=402

system indeed added to the excitement, but also uncertainties of sustaining the enviable performance of the industry.

4.2.1 Forces motivating and acknowledging demand

From an economic standpoint, the demand for casino tourism as a luxury is largely determined by the change in income of the patrons, but their preferences in choosing different games and the different budgets that they are prepared to spend on gaming and non-gaming entertainment activities for a trip are actually influenced by a set of socio-economic factors as discussed in Chapter 3 (e.g., the "peer and neighborhood effects" on the enthusiasm of Chinese patrons towards baccarat). Besides, following the rapid economic growth and development of China, increasing income and wealth have driven the mainland Chinese to spend on luxuries (including tourism), which has been a clear trend in the twenty-first century. Given that the propensity to gamble is a Chinese culture but gambling is prohibited in mainland China, the development of Macao as a casino tourism destination unarguably attracts Chinese outbound tourists.

To the Chinese patrons who are experiencing a rise in their income and social status, casino gaming and a wide range of non-gaming hospitality and entertainment services provided by the IRs not only meet their demand but also stimulate their latent demand for casino tourism. For example, an empirical study about the motivations of Chinese tourists visiting Macao finds that their "casino excursions are primarily motivated by five factors: entertainment and novelty seeking, leisure activity, escape from pressure, casino sightseeing, and socialization." (Wong and Rosenbaum 2012, 32) Given the propensity of the mainland Chinese to spend on casino tourism, the supporting policies of the Chinese government towards the tourism development of Macao (and Hong Kong as well) are actually conducive to the spending of Chinese patrons in Macao. Since 2003, the Chinese government has introduced an Individual Visit Scheme (IVS) which relaxes the visa quota and simplifies the visa application procedure for the mainland Chinese to travel to Hong Kong and Macao. Right after the enforcement of this new policy in 2003, visitor arrivals from mainland China surpassed that to Hong Kong and became the largest source of visitor arrivals to Macao thereafter (as shown in Figure 4.2).

In addition to the socio-economic and political forces, increasing demand by the mainland Chinese for casino tourism in Macao during 2005 and 2019 was also distinctly driven by some particular unorthodox and irregular individual behavior. Specifically, owing to funds and capital flow controls by the Chinese government, casino gaming in Macao has long been a channel for funds and capital flight activities undertaken by high-income individuals and even local governments outside Mainland China. Parallel to the rapid expansion in the scale of casino tourism in Macao, the successful transformation/reorganization of a large share of the third-party operated VIP gambling rooms (which were operated by junket operators in the grey area of the regulatory system) actually promoted the demand for casino gaming as a channel for funds and capital flight outside Mainland China. As Gunter (2017, 113) indicates, a "method of suspected capital flight involves gambling junkets to Macau that increase sharply beginning in 2002" (Gunter 2017, 113; also see McCartney (2015, 531 – 532)).

4.2.2 Dynamic and controversial supply-side factors

Given the growing and potential demand as motivated by the SCPE forces, it was evident that the IR operators effectively unleashed their entrepreneurship to develop comprehensive non-gaming attractions to obtain both gaming and non-gaming revenues from the Chinese patrons. These not only reflected the development of various glorious iconic properties, but most importantly, the high quality of services offered to the patrons. From the opening of the large-scale IR in the Venetian in 2007 to the considerable investment made by other IR operators (Galaxy, Melco, MGM, SJM and Wynn) on a strip of reclaimed land named Cotai, the dubbed "Cotai Strip" in Macao has been crowned the mini Las Vegas Strip in Asia. Together with other new properties that emerged all around Macao, high-end retail stores, Michelin star restaurants and world-class shows and performances, etc., catered to the enjoyment of patrons in addition to their casino gaming activities. In contrast to the STDM landmark, Hotel Lisboa, all of the new IR and hotel properties that opened during 2006 to 2019 definitely offer unprecedented dynamics in the development of the industry.

Due to the aforementioned unorthodox motivations of certain mainland Chinese for casino tourism, a group of highly independent junket operators (gaming agents) with their particular social networks and connections were also seen trying to find ways to further expand their businesses by establishing a kind of individually operated VIP gambling room within the IRs and other casino hotels (Ho 2022; Siu 2023). To a large extent, the dramatic growth in the reported GGR of the industry between 2005 and 2013 (see Figure 4.1) was indeed derived from these third-party operated VIP gambling rooms, which was reflected in the official published statistics of the gaming revenue from VIP Baccarat. As exhibited in Figure 4.3, the third-party operated business model contributed to 60% *to* 70% of the reported GGR before 2014. Nevertheless, owing to the controversial nature and ambiguity of the legitimacy of this business model (which will be explicated in the following sections), its sustainability would easily be affected if there are any changes in the political interests and regulatory requirements.

In view of the possible acceleration in the amount of funds leaving the country and capital flight, and the ever increasing extravagant spending of the community, the Chinese government took serious anti-corruption and anti-extravagant actions in 2014 to deal with these issues. As a SAR of China, the MSAR government also acted accordingly. Thus, these changes in the political interests at the outset clearly led to an extensive adjustment of the industry between 2014 and 2016 (see changes shown in Figures 4.1 and 4.3). As an integrated part of casino tourism, the non-gaming businesses of the IRs also experienced obvious downturns during this period of time. Hereupon, change in the development path of casino tourism in Macao became inevitable (as reflected in the decline of VIP baccarat and increase in mass-market turnover shown in Figure 4.3).

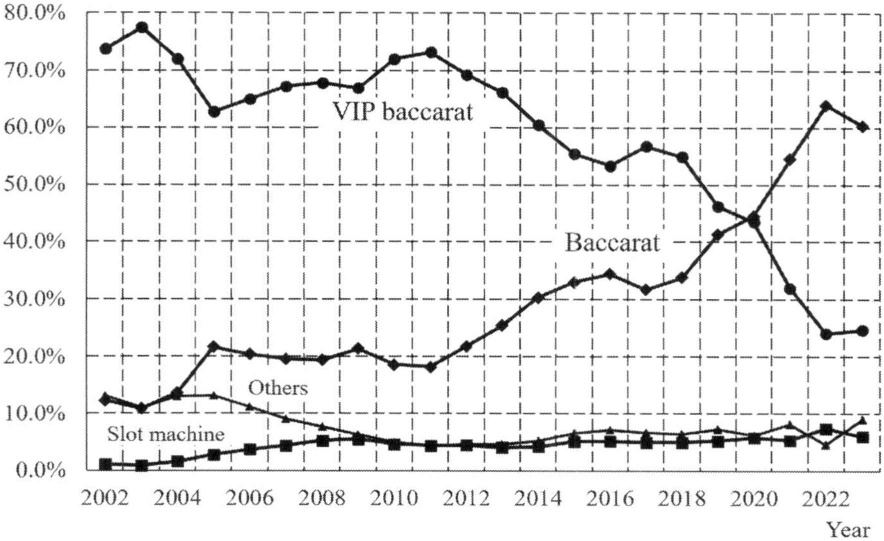

Figure 4.3 Contribution of different gaming businesses to gross gaming revenue of Macao (2002–2023)

Source: Author (measured with data derived from the Gaming Inspection and Coordination Bureau (DICJ) website of the MSAR government: https://www.dicj.gov.mo/web/en/information/index.html)

4.2.3 A lagging and loosely regulated legal system

The legal system in Macao was administered by the Macao-Portuguese government for centuries, which was "largely guided by principles of the civil law system rather than prescriptive rules" (Eadington and Siu 2007, 13) like those in the common law system. Besides, considering the social and economic realities (i.e., a Chinese-based society in which economic income highly relied on unorthodox businesses like gambling), the comprehensiveness and enforcement of laws for businesses had long been deficient (otherwise, those businesses could hardly continue and expand). Under these particular contextual settings, the MSAR government promulgated a new version of its gaming law in 2001 (hereinafter "Gaming Law 2001") to replace the monopoly of its casino gaming with an oligopoly, and a new law and a new regulation in 2002 and 2004 respectively to guide the organization and some important business practices like casino credit of the gaming agents. Nevertheless, the contents of these laws and regulations actually lagged behind the rapid transformation of the business organization/routines of the gaming agents, and were rarely enforced.

The enviable GGR growth found before 2014 (especially the surge between 2009 and 2013 as shown in Figure 4.1) was mainly driven by the rapid expansion in the scale of the VIP gambling-room and related business. However, the vitality of this business segment was supported by an increasing number of questionable business activities operating in the grey area of a lagging and loosely regulated legal system. Therefore, many gaming agents could reach out and bring in the "high income (VIP) group in Mainland China to take advantage of their fondness for gambling and extravagant lifestyle, as well as facilitating questionable/illegal flows of funds" (Siu 2023, 333) outside the country.

Owing to the nature of casino tourism (especially casino gaming), the sector is easily associated with various underground economic activities and socially/morally controversial activities. Therefore, a lagging and loosely regulated system provides the environment that promotes these activities, and hence contributes to the business turnovers of the industry. Evidence from Macao verifies that the contents of a legal system and enforcement of the related laws and regulations are negatively correlated with the short-term business turnovers of casino tourism, especially casino gaming.

4.3 Macao model dilemma

Between the mid-1980s and 2020, some of the highly independent and powerful gaming agents obtained informal contracts (hidden accords) from a casino concessionaire (i.e., a gaming license holder) and operated their own casino gaming business within the properties (generally called the VIP gambling rooms) of the concessionaire. In practice, the gaming agents who deal with the concessionaire directly were referred to as the VIP gambling-room operators. To carry out the business, a large number of individual gaming agents at various levels of the organization hierarchy were arranged to approach the actual players (gamblers) and earn a certain amount of commission from the sales of the gambling chips. Then, the VIP gambling-room operators would share a pre-determined proportion of the GGR derived from their business with the concessionaire. Over a long period of time, the practices of these third-party operated VIP gambling rooms became highly associated with irregular/illegal business activities which were practicing in a grey (or lawless) area in the legal system (Leong 2002; Siu 2006). Basically, this business organization constituted the "Macao model," which was quite different from the organization of the junket operators (gaming agents) in the Western casino tourism destinations like Las Vegas.

Even though the specific organization and business routines of the Macao model transformed over time (i.e., between the 1980s-1990s and 2000 to 2020) in congruence with the changes in the social and regulatory environments, the underlying issues were long term. The lack of legal oversight (or operating in a lawless area) and the pursuit of the various parties of the monetary benefits associated with casino gaming led to negative social impacts and various regulatory challenges.

4.3.1 Origins and drivers of the model

Aside from the adverse social impacts and regulatory challenges under the practices of the third-party operated VIP gambling rooms, the introduction of this business model by the STDM in the mid-1980s in fact represented its innovative reaction to the revision of Gaming Law 1961 by the Macao-Portuguese government in 1982 (hereinafter "Gaming Law 1982"), which expressed that the government might consider issuing a maximum of three gaming licenses instead of a single one. Nevertheless, another legal article stated the incumbent (i.e., the STDM) might still retain its monopoly license with approval by the Macao-Portuguese governor. Since gaming taxes were likely a major consideration of the government in the extension of the existing monopoly license, a workable proposal to retain the monopoly position of the STDM would be to increase its GGR and hence pay more gaming taxes. To this, contracting out part of its gaming business informally to any third party who was able to bring in patrons and receive some of the GGR from their business with the STDM would be an ideal model. And indeed, it was (Liu 2002, 406 – 415; Siu 2006).

Generally speaking, three types of benefits could be derived from the STDM under the particular social and political environments in Macao from 2000 to 2020 (Siu 2006, 972 – 974). First, to obtain the rights to operate the VIP gambling rooms within the properties of the STDM, the third parties had to underwrite (informally) a certain amount, usually a large number, of gaming chips (called "dead chips" in Chinese or non-negotiable chips) periodically which had to be played and could not be exchanged back into cash. The transactions would not be settled until the chips were lost on the gaming tables. Thus, "a stable and sizable amount of gaming revenue was secured with the least amount of effort required" (ibid., 972) by the STDM. Second, although a casino license holder was legally prohibited from granting casino credit to its patrons during that period of time, related activities that could be undertaken by other third parties (i.e., VIP-room operators and their associated gaming agents) were not spelled out legally (i.e., a grey area). Therefore, the high-end patrons could be approached by the gambling-room operators and their associates through the provision of gaming credits. Third, as a result of the two arrangements above, the STDM could ensure a large inflow of GGR without the risk of bad debt from granting casino credit by itself.

Consequently, as a result of the increase in the gaming "tax rates between 1987 and 1992, the gross tax payment made by the monopolist to the government recorded a 30 percent to 40 percent annual growth" (ibid., 973). Meanwhile, the STDM renewed its monopoly license successfully until 2001.

4.3.2 Transformation of the business routine in a new era

Although the third-party operated VIP gambling-room model provided the expected monetary return to the STDM and hence enhanced its ability to contribute to the tax revenue of the government and social welfare of the local community, their practices were actually dominated by a few powerful triad groups. Practicing in a largely lawless environment, competition between these triad groups for coveted economic benefits associated with casino gaming business not only led to serious social chaos like gun shootings, kidnappings, arson, etc., but also explicitly threatened the regulatory power of the local police force in the second half of the 1990s (Callick 1998; Leong 2002; Richburg 1999; Segal 1998; Siu 2006). In view of the establishment of the MSAR at the end of 1999, serious actions were then seen taken by the Macao police force and related law enforcement units after 1998 to resolve the problems.

Despite the anxiety of the general public about whether the newly established MSAR government could resolve the problems, social stability and order were clearly reinstated with the full support of the Chinese government. Besides, the MSAR government also made a decisive choice to replace the monopoly with an oligopoly in 2002 through international tendering. In congruence with the potential changes in the industrial structure, revised and new laws and regulations were also promulgated in the attempt to better regulate and supervise the redevelopment of casino tourism in Macao. Other than the Gaming Law 2001, a new law to stipulate the qualifications and rules for the business practices of the gaming agents, and another one to regulate the practices of casino credit were also promulgated in 2002 and 2004 respectively.

Right after these changes and the promulgation of these new laws and regulations, it was evident that the traditional third-party operated business segment slowed down in 2004 and 2005 (see decline shown in Figure 4.3). Nevertheless, the gaming agents were still attracted by the evolving industry and potential monetary benefits, so they also started to transform their business routines and test the tolerance of the government in terms of its commitment to the enforcement of the new laws and regulations, as well as the possibility of reaping benefits from the loopholes (i.e., grey areas) in the legal documents. Indeed, the gaming agents found ways to transform their business routines and increase their business scale.

Facing a much larger potential market from mainland China in the new millennium, it appeared that the gaming agents did not need to compete for relatively small populations of patrons from other countries/regions like Hong Kong, Taiwan and Japan as in the last two decades of the twentieth century. Besides, parallel to the significant improvement in the leisure and entertainment facilities provided by the IRs, the gaming agents largely approached the wealthy mainland Chinese (including government officials) by offering high-quality services to them to elevate their enjoyment during their casino tourism trips in Macao. In the process, however, irregular business activities like usury, shadow banking and money laundering also increased, and most critically, helping funds and capital flight from mainland China were suspected (and identified in many cases) as new business routines carried out by the third-party operated VIP gambling-room operators and the related gaming agents.

In terms of these phenomena, it is evident that although the practices of the Macao model no longer created serious chaos to Macao society but in turn brought in a considerable amount of high-class "VIP" business turnovers for the industry (and hence contributed largely to the tax income of the government as compared to the STDM era), their business activities indeed involved "**V**ery **I**nappropriate **P**ractices." This change actually shifted the core of the problems to other areas, which were mainly social in nature, and "exported" to mainland China. This increased the challenges to the Chinese government who had to enforce some of its important laws and regulations, especially those that control capital flight and corruption.

4.4 Reforms and prospects

The development of the IRs in 2000 to 2020 created large-scale properties with alluring exterior and interior designs (i.e., the fixtures) which have unarguably established a solid base for Macao as an attractive and competitive casino tourism destination. In economics, the advantages of both the internal economies of scale to the individual IR operators and external economies of scale to the industry as a whole are found to a large extent. For example, over ten IRs with unique features were launched respectively by the six IR operators on the Cotai Strip. These facilities have increased the popularity of the Cotai Strip for tourists not only for gaming, but for leisure and entertainment activities.

Nevertheless, the downsides associated with the third-party-operated VIP gambling-room model continued to create uncertainty and threaten the sustained growth of the industry. Actually, extensive changes made between 2014 and 2016 (as shown in Figure 4.1) caused by the changes in political interests/orientation, thus tightening of the regulatory policies, set the stage for the need to reform the business organization and routines of the casino gaming and its interrelationship with the non-gaming tourism sectors (McCartney 2015). Despite the continuous efforts made by the beneficial groups (especially the VIP gambling-room operators and their associates) to preserve the existing business model, the MSAR government was determined to revise and improve the organization of this key industry, which did not leave much space as historically allowed for the further transformation and resurgence of this socially and politically controversial third-party-operated VIP business segment. Coincidently, the outbreak of the COVID-19 pandemic in 2020 indeed provided an opportunity for the MSAR government to sit back (when the majority of the casino tourism businesses had to shut down or partially close) and conduct detailed and fundamental revisions of its gaming laws and regulations to ensure the sustainability of this industry in the long run.

4.4.1 Forces and actions that reconstruct the industrial organization

As compared to the 1980s and 1990s, the specific social and regulatory problems were also bred by the practices of the third-party-operated business segment (especially the VIP gambling rooms), although the problems produced from the practices of casino gaming in 2000 to 2020 were not the same (as discussed in Sections 4.3.1 and 4.3.2, respectively). As aforementioned, social chaos in the late STDM era was largely caused by the competition between the major gambling-room operators (who were at the same time key members of some of the powerful triad groups) to pursue the casino benefits. These problems were indeed exacerbated by the absence of a needed regulatory system to curb the related activities. Given the full support of the Chinese government to the MSAR government in the new millennium, social stability was finally restored. Nevertheless, new problems transpired in different forms when there were still opportunities that allowed some of the third parties to pursue the monetary benefits associated with the operation of casino gaming in the particular environment of Macao.

New challenges and forces

As aforementioned, although the MSAR government promulgated new laws and regulations following the restructuring of the industry in an attempt to regulate the business activities of the gaming agents in the third-party-operated VIP gambling rooms, their contents actually lacked some needed details for their effective enforcement. Thus, a group of new VIP gambling-room operators and related gaming agents successfully found ways to transform their business routines and revitalize their businesses in the gray areas of the legal system again. Consequently, since unorthodox and irregular business activities associated with the VIP gambling rooms were not checked in a timely manner, so socially and politically undesirable outcomes snowballed again, although the specifics of these problems (e.g., extravagant spending, family problems, corruption, usury, capital flight etc.) were different from those of the 1990s. Therefore, counterforces to these new problems were also generated.

Other than the social and political concerns, the dramatic growth and development of casino tourism mainly in the large-scale IRs and catering to the VIP segment from 2000 to 2019 also showed that some needed development in other parts of the economy were limited or even crowded out (especially the local small and medium enterprises (SMEs)). Although it had long been the public interest to diversify the industrial structure of the local economy moderately through the development of casino tourism, its progress evidently led to a more monotonic structure of the economy. Besides, under the formal proposal promulgated by the Chinese government in 2019 to construct the Guangdong–Hong Kong–Macao Greater Bay Area (GBA) as part of a larger project in the Belt and Road Initiative (BRI), the long-term path of development would have to be revisited according to these new national goals at the outset.

Also, following the surge in the VIP gaming segment, the dependence of other tourism sections like catering and retail on this segment was mentioned, while the SMEs appeared to be in poor predicament (or crowded out) as business costs (especially labor costs and rent) soared. This challenge is actually reflected in the low and falling tourism ratios[5] for the related business sectors. For example, the tourism ratio of the "hotel and similar establishments" of Macao fell from 36.6% *in* 2010 *to* 28.6% in 2015 and further *to* 25.4% in 2019.[6] Although this ratio increased in some of the other sectors like retail and food and beverage after 2014, these sectors continued to greatly rely on casino gaming.

Decisions and actions

In view of the changing macro-environment and expiration of the existing gaming licenses in 2022, it was observed that a series of actions was undertaken by the MSAR government between 2019 and 2022 to review and amend its gaming laws and related regulations for launching radical changes in the industry when tenders were to be invited for the new gaming licenses starting in 2023. As reported by some of the local media, the MSAR government started to review Gaming Law 2001 in 2019. However, owing to the unexpected outbreak of the COVID-19 pandemic in 2020, the task was delayed. Finally, following a public consultation in 2021, a new version of the gaming law was promulgated in 2022 (hereinafter "Gaming Law 2022"[7]). In comparison to the first three versions of the gaming laws (i.e., Gaming Laws 1961, 1982 and 2001), it is obvious that Gaming Law 2022 attempts to close most, if not all, of the loopholes of the previous versions of the laws. To improve/ensure the enforceability of Gaming Law 2022, a new law which focuses on the business practices of the various parties was also promulgated in 2022 (hereinafter "Business Regime 2022"[8]).

For example, Gaming Law 2022 and Business Regime 2022 state clearly that a gaming agent may only be allowed to provide agency services to one casino gaming license holder by receiving a pre-approved rate of commission from the related business turnover. Besides, any form of individually operated gaming business within the property of a gaming license holder and sharing of the GGR is prohibited under a new gaming license that commences in 2023. In addition, while detailed business transaction records between a gaming license holder and its gaming agents have to be filed and submitted periodically to the regulatory authorities (i.e., the Gaming Inspection and Coordination Breau (DICJ) of the MSAR government) for monitoring purposes, a gaming license holder is expected to assume joint responsibility with its associated gaming agents if illegal activities are found by the latter (for a detailed discussion, see Siu (2023)).

Also, to clearly convey to the market that the government is determined to prevent any possible changes of the third-party-operated VIP gambling rooms and the resurgence of irregular/illegal business activities as historically found, the MSAR government arrested two gambling-room tycoons in 2021 and 2022 respectively (see, e.g., Costa (2022) and Smith (2021)). In other words, the decades-long third-party-operated gambling-room model that operated within a gray area of the legal system really came to an end under the provisions of these new laws.

4.4.2 Developing a sustainable path

On top of resolving the most challenging problem of the third-party-operated business segment, the MSAR government also highlighted two other areas for the industry to ensure the sustainability of its redevelopment and its real and long-term contribution to the local economy. These include the explicit requirements for the industry to emphasize the development of the mass market (instead of the VIP market) and the interplay between the development of the IRs and the local tourism sectors.

In terms of mass market development, more middle-end instead of high-end non-gaming tourism facilities are expected to be further developed. According to the ten-year new contracts signed by the six IR operators with the MSAR government at the end of 2022, over USD12.5 billion were to be committed for the development of non-gaming facilities from 2023 (Plataforma 2022). Besides, new sources of patrons would also likely be sought. As a motivation, Gaming Law 2022 states that 5% of the required contribution from the GGR of the casino concessionaires to Social and Cultural Affairs of the MSAR government could be exempted if that part of the GGR is derived from foreign patrons (i.e., beyond mainland China, Hong Kong, Macao and Taiwan).

To realize the interplay between the development of the IRs and the local tourism sectors, all IRs have expressed their willingness to participate respectively in the various revitalization projects in the old town areas of Macao. Indeed, the old town areas have rich local cultural and heritage resources that can be part of the tourism appeal of Macao. To a large extent, participation of the IRs in these projects not only ensures the needed capital investment for the infrastructure projects to improve the efficiency of the flow of visitors but also helps to increase the economic returns of those resources by providing modern management advice to the local SMEs that are operating in those areas.

On the other hand, it is worth noting that as compared to the historical periods, "the MSAR government is assuming an active role in this new era to ensure the implementation of these plans" (Siu 2023, 340) and the effective enforcement of the new laws and regulations. Although all these radical adjustments to the business organizations and routines do not mean that the GGR will return to the pre-COVID levels in 2019, never mind the GGR of unprecedented heights in 2013 (see Figure 4.1), a more sustainable path for the development of casino gaming in Macao's casino tourism is really enroute.

4.5 Summary

To the development of Macao's casino tourism, a separate monograph can indeed be produced. In this chapter, the SCPE forces and major factors that have framed the path of development of Macao from a monotonic gambling hub in the nineteenth century to a world-renowned casino tourism destination in the twenty-first century are explicated. Besides, the unique dynamics (particularly the practices of the third-party-operated VIP gambling rooms and the related gaming agent business activities) that have contributed to the modern development and dramatic performance of this industry, on the one hand, but limited the sustainability of the Macao model, on the other hand, are elucidated.

From 2000 to 2019, the success and dramatic performance of casino tourism in Macao unarguably conveyed crucial messages to other existing and potential destinations in the ESEA. While some destinations might attempt to learn and borrow from the Macao model, others might remain cautious about the downsides of this model, especially the likely social and political/regulatory issues. Nevertheless, the radical decisions of the MSAR government in 2022 to restructure the business organization and redirect the business routines of the industry by removing the third-party-operated business segment and emphasizing mass market development not only reconstructed the frontiers of the development of casino tourism in Macao but also had explicit impacts on the progress of some of the other ESEA destinations.

As for the changes in the progress of Macao's casino tourism in 2023, these indeed provide a critical message that the tolerance of the Chinese government to the expansion of irregular/illegal business activities related to casino tourism is waning. Specifically, using casino gaming and tourism activities to channel out funds and conduct capital flight out of mainland China has been seriously addressed by the Chinese government through the amendment of Article 303 in its Criminal Law in 2021. Under this article, anyone "organizing the participation of mainland residents for gambling out of borders" (Mcura 2021) would be charged with criminal offence and imprisoned (more details are presented in Chapter 9). To this, traditional activities taken by the Macao-style gaming agents to promote casino gaming tours for IRs in mainland China are illegal.

In a nutshell, the evidence and experiences of Macao show that casino tourism processes its particular attributes to develop as a modern service sector. Nevertheless, in the social and political contexts, possible downsides associated with the practices and expansion of this industry do require the provision of an effective regulatory system and enforcement of related laws and regulations. Otherwise, a good business performance in monetary terms could be illusive and hardly sustainable in congruence with societal advancement and changes in political interests.

Notes

1 This role as a trade hub and transshipment center stopped several times between the mid-sixteenth and mid-nineteenth centuries owing to the changes in the political

interests of the Japanese and Chinese governments respectively in trade with the Western world.

2 For a short history on the trade of coolies and its relationship with the gambling business in Macao, see, for example, Corvo (n.d.). On the other hand, the vaeseng lottery is the earliest form of a Chinese lottery operating not just in Macao but also in mainland China. For discussions about pacapio and the vaeseng lottery, see, for example, Pina-Cabral (2002, 90–100).

3 Geographically, the urban development of Macao is highly restricted by its small land area. Thus, investment for the reclamation of land in the coastal area has been an essential measure to overcome this limitation. To a large extent, the operations of the gaming industry provided the financial resources for the land reclamation of Macao over time. Therefrom, the land area of Macao was increased from 11.6 km^2 in 1912 to 15.6 km^2 in 1980, and greatly increased to 25.4 km^2 in 2000 and further more so to 32.9 km^2 in 2020.

4 In 2008, the definition adopted by the DSEC to record and report the number of visitor arrivals was revised to align with international standards. Mainly, transit passengers were excluded from the statistics thereafter. Thus, the total number of visitor arrivals from different sources after the adjustment was lower, which is exhibited by the broken lines in 2008.

5 A tourism ratio is a statistic to show the direct spending made by tourists in a tourism sector like hotels, net of any transfer payment from other sectors. For example, complimentary hotel rooms and food and beverage provided by casinos to their patrons do not count as direct spending by tourists in these accommodation services. In other words, a larger tourism ratio means that a sector is more independent in its operations and how it derives its business income.

6 Data derived from the Government of Macao Special Administrative Region Statistics and Census Service (DSEC) website: https://www.dsec.gov.mo/en-US/Statistic?id=408.

7 Law No. 7/2022 (in Chinese). Available at https://images.io.gov.mo//bo/i/2022/29/despce-121–2022.pdf. Accessed July 19, 2022.

8 Law No. 16/2022 (in Chinese). Available at https://images.io.gov.mo//bo/i/2001/39/lei-16–2001.pdf. Accessed December 30, 2022.

References

Brady, Diane. 1998. "Mayhem in Macau: The Joint Is Jumpin' at the Hotel Lisboa." *The Wall Street Journal* (May 29, 1998): 1.

Callick, Rowan. 1998. "Rambo Gets Jump on Broken Tooth." *Financial Review* (May 5, 1998). Available at https://www.afr.com/world/rambo-gets-jump-on-broken-tooth-19980505-jlbha. Accessed June 13, 2021.

Corvo, Andrade. n.d. "Macau in the XIX Century: Coolie Emigration." Available at http://www.icm.gov.mo/rc/viewer/20007/866. Accessed July 4, 2020.

Costa, Sofia Horta E. 2022. "Macau Arrests Second Junket Boss on Money Laundering Charges." *Bloomberg.com* (January 30, 2022). Available at https://www.bloomberg.com/news/articles/2022-01-30/macau-arrests-second-junket-boss-on-money-laundering-charges#xj4y7vzkg. Accessed February 14, 2022.

Eadington, William R. and Ricardo C. S. Siu. 2007. "Between Law and Custom – Examining the Interaction between Legislative Change and the Evolution of Macao's Casino Industry." *International Gambling Studies* 7 (1): 1–28.

Godinho, Jorge. 2012. "A History of Games of Chance in Macau: Part 1 – Introduction." Gaming Law Review: Economics, *Regulation, Compliance, and Policy* 16 (10): 552–556.

Godinho, Jorge. 2013. "A History of Games of Chance in Macau: Part 2 – The Foundation of the Macau Gaming Industry." Gaming Law Review: Economics, *Regulation, Compliance, and Policy* 17 (1): 107–116.

Godinho, Jorge. 2014. "Casino Gaming in Macau: Evolution, Regulation and Challenges." *UNLV Gaming Law Journal* 5 (1): 1–26.

Gunter, Frank R. 2017. "Corruption, Cost, and Family: Chinese Capital Flight, 1984–2014." *China Economic Review* 43: 105–117.

Ho, Hong-Wai. 2022. "Gaming the System? The Case of Satellite Casinos in Macau." Gaming Law Review: Economics, *Regulation, Compliance, and Policy* 26 (5): 269–281.

Leong, Angela V. M. 2002. "The 'Bate-Ficha' Business and Triads in Macau Casinos." *Queensland University of Technology Law and Justice Journal* 2 (1): 83–97.

Liu, Pinliang. 2002. *Macao Gaming Industry* (in Chinese). Hong Kong, China: Joint Publishing (H.K.) Co., Ltd.

McCartney, Glenn. 2015. "When Eggs in One Basket All Cracked: Addressing the Downturn in Macau's Casino and VIP Junket System." Gaming Law Review: Economics, *Regulation, Compliance, and Policy* 19 (7): 527–537.

Moura, Nelson. 2021. "Gaming Operators, Junkets, and Private Citizens Should Carefully Study New Mainland Penal Code Amendments – Report." *Macau News Agency* (November 30, 2021). Available at https://www.macaubusiness.com/gaming-operators-junkets-and-private-citizens-should-carefully-study-new-mainland-penal-code-amendments-report/. Accessed December 14, 2021.

Pascoal, Carlos. 1995. "Macau: Brief Summary of Gambling." *Casino Chip and Token News* 8 (4): 30–35. Available at http://www.ccgtcc-ccn.com/vol8_4.pdf. Accessed June 26, 2016.

Pina-Cabral, João de. 2002. "Paradoxes: Gambling and the Imperial Civil Service Examination." In *Between China and Europe: Person, Culture and Emotion in Macao*. Edited by Pina-Cabral, João de, 79–103. New York, NY: Continuum.

Plataforma. 2022. "Gaming in Macau – Millions to Surpass Las Vegas." (December 23, 2022). Available at https://www.plataformamedia.com/en/2022/12/23/gaming-in-macau-millions-to-surpass-las-vegas/. Accessed December 28, 2022.

Richburg, Keith B. 1999. "Revenge with a Vengeance." *The Washington Post* (January 13, 1999). Available at https://www.washingtonpost.com/archive/politics/1999/01/13/revenge-with-a-vengeance/35ac9161-ed06-46e2-9ebc-064b0e6542fe/. Accessed December 14, 2000.

Segal, Philip. 1998. "Wild-West Macau Is Hoping to Hang Up Its Guns." *The New York Times* (April 3, 1998). Available at https://www.nytimes.com/1998/04/03/business/worldbusiness/IHT-wildwest-macau-is-hoping-to-hang-up-its-guns.html. Accessed January 8, 2023.

Siu, Ricardo C. S. 2006. "Evolution of Macao's Casino Industry from Monopoly to Oligopoly – Social and Economic Reconsideration." *Journal of Economic Issues* 40 (4): 967–990.

Siu, Ricardo C. S. 2023. "Back to the Future: Constructing Macao as a World-Class Casino Tourism Destination under New Gaming Laws." Gaming Law Review: Economics, *Regulation, Compliance, and Policy* 27 (7): 326–342.

Smith, Michael. 2021. "Macau Police Arrest Junket Boss in Latest Blow to Gaming Enclave." *Financial Review* (November 29, 2021). Available at https://www.afr.com/world/asia/macau-police-arrest-junket-boss-in-latest-blow-to-gaming-enclave-20211129-p59d6d. Accessed February 14, 2022.

Wills Jr., John E. and Paul Van Dyke. 2000. "Strange Shores: 442 Years of Anomaly in Macau, and Counting . . ." *Harvard Asia Pacific Review* (Summer 2000): 1.

Wong, Anthony Ipkin and Mark S. Rosenbaum. 2012. "Beyond Hardcore Gambling: Understanding Why Mainland Chinese Visit Casinos in Macau." *Journal of Hospitality &* Tourism Research 36 (1): 32–51.

Wong, Anthony Ipkin and Ricardo Chi Sen Siu. 2015. "Casino Tourism: How Does Entrepreneurship Change Macao?" In *Entrepreneurship in Small Island States and Territories*. Edited by Bodfrey Baldacchino, 218–234. New York, NY: Routledge.

Wu, Anise M. S. and Joseph T. F. Lau. 2015. "Gambling in China: Socio-Historical Evolution and Current Challenges." *Addiction* 110 (2): 210–216.

Additional readings

Cohen, Muhammad. 2017. "Beating the Odds, Changing the Game. *Inside Asian Gaming* (August 2017): 22–27.

Greenwood, Verity A. and Larry Dwyer. 2018. "Reinventing Macau Tourism: Gambling on Creativity?" *Current Issues in Tourism* 20 (6): 580–602.

He, Miao and Ricardo C. S. Siu. 2018. "Reassessing Economic Success: More Than a Decade After Casino Liberation in Macau." In *Macau and the Casino Complex*. Edited by Natalia Echeverri, Kah-Wee Lee and Stephen Al, 88–98. Reno, NV: University of Nevada Press.

Porter, Jonathan. 1996. *Macau,* The Imaginary City: Culture and Society*, 1557 to the Present*. Boulder, CO: Westview Press.

Mc Cartney, Glenn. 2014. "With or Without You? Building a Case for Further MICE (Meetings, Incentive Travel, Conventions, Exhibitions) Development in Macao." *Journal of Convention &* Event Tourism 15 (3): 173–197.

Siu, Ricardo C. S. 2007. "Formal Rules, Informal Constraints, and Industrial Evolution: The Case of the Junket Operator Regulation and the Transition of Macao's Casino Business." *UNLV Gaming Research and Review Journal* 11 (2): 49–62.

Siu, Ricardo C. S. and William R. Eadington. 2009. "Table Games or Slots? Competition, Evolution, and Game preference in Macao's Casino Market. *Journal of Gambling Business and Economics* 3 (1): 41–63.

Zandonai, Sheyla S. 2015. "Casino Development and Urban Transformation in Macau." *The Newsletter*, no. 72 (Autumn 2015). Available at https://www.iias.asia/sites/iias/files/nwl_article/2019-05/IIAS_NL72_1011.pdf. Accessed June 13, 2021.

5 IR model of Singapore and implications

As "a politically censorious and highly-regulated society" (Lee 2005, 74) in which purity and virtue are commonly regarded as part of the national goals for development, the decision of the Singapore government to approve the development of two IRs in 2005 (Ministry of Trade and Industry 2012; Szep 2005; Yap 2007) by including the business segment of casino gaming led to intense social debates (see, e.g., Chia (2007); Henderson (2006); and Loke (2004)). Nevertheless, the actions taken by the government in the process to facilitate the development and monitor the practices of the IRs in the 2010s not only turned a new page in the development of casino tourism globally but also provided valuable references for some of the existing jurisdictions in the ESEA to reconsider and improve the competitiveness and sustainability of their existing business models in the long run. In addition, the practices of this industry in Singapore offer guidelines for those that are potential casino gaming sites to assess whether and how casino tourism can be introduced in their country (see, e.g., Lee (2017) and Leu and Ko (2012)).

Despite the fact that Singapore is a latecomer to casino tourism in the ESEA, proactive public planning and government efficiency in the enforcement of the related laws and regulations actually contributed to the success of the Singapore model, which does not simply emphasize the monetary performance (i.e., GGR) of the industry but its sustainability and long-term contributions to the attractions/competitiveness of a tourism destination. In this chapter, the particular contextual settings and forces that have motivated the Singapore government to approve the development of casino tourism in the country are scrutinized in the first section. Then, the nuances of the IRs based on public preference and the role of law enforcement in the practices of casino tourism in the country are explicated in the second and third sections, respectively. Consequently, the performance of the IRs in Singapore and the implications for the future development of casino tourism in the ESEA are presented in the fourth section.

Takeaways from this chapter:

1. The decision of the Singapore government to legalize/incorporate casino gaming as a formal business sector in its tourism plans represents a cautious and strategic trade-off between the interests of different stakeholders, and the potential societal pros and cons.

DOI: 10.4324/9781003279884-5

2. The Singapore government has adopted IRs as an innovative means of achieving multiple objectives to sustain the competitive edge of its economy, while casino gaming only represents a small part of a much bigger picture.
3. Under the contextual settings in Singapore, IRs and IR operators also need to refer to the guiding principles of "Integrity and Responsibility."
4. To ensure the potential contributions of the IRs to tourism in Singapore, and minimize the potential adverse impacts of casino gaming, the government closely supervises the practices of casino tourism through tough laws and regulations.
5. As the rules of the game for businesses are clearly specified, IR operators well-exhibit their entrepreneurial capability and experiences to choose the right market segments and promote their business strategically, thus contributing to the successful development and sustained growth of casino tourism in Singapore.
6. The formal decision of the Singapore government to retain the duopoly structure of the industry in 2019 and extend the licenses of two IRs to 2030 shows its leadership capability and ability to plan proactively for the long-term development of the industry.
7. Empirically, a core feature of the "Singapore model" is the emphasis on social safeguards (i.e., to minimize the adverse social impacts of casino gaming although it is difficult to entirely eliminate the related issues).
8. The Singapore model provides valuable references to other existing (or potential) casino tourism jurisdictions in the ESEA so that they can improve (or search for) the appropriate path of development which would accommodate the contextual settings of their own society rather than simply reproduce the fixtures of IRs.

5.1 Cautious and strategic public decisions

Although casino gaming was outlawed in Singapore before 2005, non-casino gambling activities available to Singaporeans like lotteries (e.g., the Singapore Pools), horse racing and sports betting in the country had long been socially controversial issues (Chia n.d.; Winslow et al. 2015). Actually, the gambling culture of the masses in Singapore is largely linked to its largest ethnic group – the Chinese[1] "who are well integrated into the community to engage in more gambling activities" (Winslow 2015, 1384). These also include their engagement in casino gaming outside the country. On the other hand, as the Muslims (the second largest ethnic group in Singapore) prohibit gambling in deference to their Islamic faith, the decision of the Singapore government to legalize casino gaming as a 24/7 (i.e., operated continuously 24 hours a day and 7 days a week) gambling business definitely raised many different social concerns and challenges. Thus, to ensure that the development of casino tourism would really contribute to the long-term benefits of the local economy, especially tourism, the IR operators are strictly required to comply with the related laws and regulations when conducting their businesses.

5.1.1 IR as an innovative service with multiple objectives

Considering the possible negative social impacts that range from problem and pathological gambling and the opposing forces from the different ethnic and social groups, it is clear that the Singapore government had to be cautious in balancing the interests of the different stakeholders. Otherwise, the success of their proposal to develop casino tourism and its potential contributions to the economy would not be realized, and social and political stability would also be threatened. Amid the controversies in the communities, the Singapore government demonstrated its talent and political power to formulate a particular path for the development of this industry. The plan was largely framed by the innovative application and formal recognition of IRs, which included but at the same time downplayed the business segment of casino gaming. As the government emphasized in its formal statement of approving IRs in 2005: "an IR is not just a casino. An IR is one significant idea we must consider, that will help us reinvent Singapore" (Ministry of Trade and Industry 2005a). In other words, IRs are different from casinos or casino resorts, and "they should be called leisure, entertainment and business zones" (ibid.).

Despite that the operations of mega casino resorts are considered to be an integration of casino "gaming into the hotel, retail, restaurant, convention, entertainment, and other components of these immensely complex developments" (Eadington and Christiansen 2007, 228) in the experience of Las Vegas, the notion of IRs as formally stated in the legal document to include the business segment of casino gaming was presented for the first time in Singapore. In practice, although IRs are found in Europe and Australia since the 1970s, a wide range of hospitality and entertainment facilities are integrated with various key tourism elements and not casino gaming in a destination for tourists. For example, "integrated ski resorts" (Pearce 1978, 142) constructed in beach areas in France and "integrated resorts" (Stanton and Aislabie 1992) developed in various tourism destinations in Australia were among the early forms of such properties introduced to boost the development of (non-casino) tourism.

Contextually speaking, the effective promotion of the concept and practices of IRs in Singapore allows the government to meet different objectives. With reference to the formal statements of the government that approved IRs (Ministry of Trade and Industry 2005a), Henderson (2006) argues that the country benefits from the development of IRs in three major areas. First, IRs would contribute to revitalizing the "unexciting" (ibid., 174) image of Singapore for tourism and reinventing the country as an attractive and vital city to individual and business visitors (especially the growing number of outbound Chinese visitors). Second, tourism development through IRs provides a promising opportunity to diversify the economy given that the traditional manufacturing industry in Singapore is declining. Third, by making attractive and internationally competitive IRs available to the locals, "[j]obs would be provided and some of the $S1.5 billion gambled by Singaporeans abroad and on cruises would be retained"[2] (ibid., 175) in Singapore.

5.1.2 Prudent arrangements and strategic planning

As aforementioned, social concerns about the adverse impacts that range from making a more convenient form of gaming activity (i.e., casino gaming as compared to lotteries and sports betting) available to the locals not only could amplify the potential gambling problems of the locals but also led to criticisms from some of the ethnic and social groups. To this, it is observed that the government maintained a low profile prior to 2005 during its preparation work before a formal announcement was made about its decision to approve the IR concept. For example, although it is evident that consensus/support for the legalization of casino gaming in the country would not likely come from the Muslim community, the government communicated proactively with them to reduce their concerns (although this was unlikely) before the formal announcement of its decision/arrangements for IRs (Ministry of Trade and Industry 2005b).

Besides, before arriving at a formal decision, the Singapore government issued a Request for Comments (RFC) at the end of 2004 (Getforme Singapore 2004) to gauge the market response. In the RFC, interested investors (potential operators) were invited to submit their suggestions as to how they would develop IRs in Singapore. Through this step, the government held the view that if the submitted "idea of IR was not viable, or that investors were only interested in opening gambling joints" (Ministry of Trade and Industry 2005a), then they would withdraw this idea. Based on the favorable feedback from the RFC (including suggestions on how to address the potential adverse impacts of the operation of casinos), the government then finalized its decision to move forward and made a Request for Proposal (RFP) in 2005 for two IR licenses in Marina Bay and Sentosa, respectively.

The government specified four particular criteria in the RFP for the evaluation of submitted proposals: "tourism appeal and contribution; architectural concept and design; quantum of development investment; strength of the consortium and partners" (Ministry of Trade and Industry 2012). Thus, new iconic IRs were anticipated instead of properties that replicated those of an existing jurisdiction like Las Vegas and Macao, which would be designed by IR operators to increase the attraction for tourism in Singapore. As expected, the winners of the two IR licenses (i.e., Las Vegas Sands and Genting International) brought in attractive new tourism images and facilities for the country. These include their well-known landmark properties – the Marina Bay Sands (MBS), with architecturally unique features which include an infinity pool on the top level of the property (see Figure 5.1), and Resorts World Sentosa (RWS), by including the only Universal Studios theme park in Southeast Asia.

To ensure the competitiveness of the industry, the government offered a low tax policy to support the development of the IRs. Thus, the gaming tax rate levied on the GGR derived from the mass gaming market is 15% (Ministry of Trade and Industry 2019), which in principle is close to the corporate tax rate in Singapore, while the rate on the premium (or VIP) market is only 5% (ibid.), which reflects the higher business cost that the IRs incur in approaching premium patrons (revision of the gaming tax rates in 2022 will be discussed in Section 5.3).

Figure 5.1 Marina Bay Sands

Source: Author's photo

On the other hand, as an attempt to reduce the likelihood of the locals from gambling in the casinos, the government imposed casino entry levies of $S100 (around USD60 to 70) for a daily pass and $S2,000 for an annual pass (Ministry of Social and Family Development 2019)[3] to Singaporeans and permanent residents in Singapore who want to enter the casinos in the IRs. Although there is much controversy over the effectiveness of this policy (Yeu et al. 2012), the will of the government to reduce the aggregate adverse impacts on the local community from casino operations in the IRs is undeniable. Actually, this policy illustrates the argument of Eadington (1998, 54) as presented in Chapter 3 that to ensure the potential contributions of casino gaming to the local economy, the objectives of the gaming laws and regulations and the potential impacts from the accessibility of the gaming facilities by locals and tourists should be one of the essential public policy questions that the government has to carefully consider according to real societal situations (contextual settings).

5.2 Industrial performance in a dynamic environment

Given the particular social concerns of the potential adverse effects due to the practices of casino gaming in the IRs, and the inclination of the government to consider the business activities of the industry within the social and political contexts of Singapore, the general expectation was that the market positioning of the IRs would align with public interests. To this, the business practices of the IRs are actually guided by the principles of "**I**ntegrity and **R**esponsibility" at the outset to meet social expectations. As the rules of the game for conducting business were clearly specified, the two IR operators demonstrated their entrepreneurial capability and leveraged on their experiences to choose the appropriate market segments and promote their business strategically. Facilitated by other external factors like positive income and wealth effects in Asia as a whole (especially China and the growth in India) and better infrastructural connections in the region, the efforts of the Singapore government and IR operators provided them with an initial payback in the 2010s. Aside from gaming revenue, non-gaming tourism spending in the country was also increasing.

5.2.1 Business practices under a highly regulated environment

Potential investors who submitted their bids for the IR licenses gained the sense that their business activities would be subjected to strict monitoring under a highly regulated system in Singapore when they considered the position that the government held in the development of the IRs and through the actual documents and announcements presented throughout the process of RFC and RFP in 2005. Besides, as a new and controversial industry in the country, continuous government efforts to finalize the related laws and regulations following the approval of the two IRs were not unanticipated.

For example, the significant but controversial role of junket operators (or gaming agents or gaming promoters) in the Asian market (particularly the Chinese market) means that their qualifications and business activities and relationships with the IR operators are highly regulated and supervised by the Casino Regulatory Authority (CRA) in Singapore (Tan 2011; GGB 2010). These mainly include how the gaming agents manage the gaming chips and credits with patrons who are solicited to visit the casinos, as well as the required detailed records of the fund flows of the premium (VIP) players. In addition, the kinds and extent of the business activities (e.g., promotion of casino gaming) that are allowed and prohibited are specified to the IR operators and the gaming agents (Tan 2011) and subject to the supervision of the CRA. Accordingly, the CRA only issued the first two junket operator licenses in 2012 (i.e., two years after the opening of the two IRs) "after a series of stringent checks which lasted more than a year" (Song 2012).

Under a highly regulated business environment, it is evident that even though the two IRs also work with gaming agents to bring in high-end (premium or VIP) patrons, especially those from mainland China, the agreements between the IR operators and gaming agents are different from those in Macao (i.e., as compared to the traditional "Macao model" before 2023 as specified in Chapter 4). In Singapore, gaming agents who work with the IRs cannot operate their own VIP gambling rooms within the properties of the IRs on top of the rigorous background checks and approval requirements. In addition, business transactions that involve the gaming agents with the patrons like fund flows and credits have to be well-documented and are subject to strict and ongoing supervision of the CRA.

Besides, to address the explicit social concerns about problem gambling issues, the two IRs declared upon the opening of the properties that they would include "responsible gaming programmes which cover staff trained in recognising and aiding those affected and the option of setting a limit on the amount to be gambled, administered by the casino" (Henderson 2012, 143). The ongoing supervision of the industry through responsible gaming programs shows that the IRs also adopted new measures to better meet their expected responsibilities. For example, Mei (2019) reports that the CRA "was working together with the casinos on measures which could also include the option of having casino patrons self-regulate their gambling habits by voluntarily setting caps on their expenditure and time spent at the tables."

5.2.2 Industrial performance from IR practices

Despite the particular (and demanding) public expectations of and requirements for the practices of IRs in the development of casino tourism in Singapore, the trademark effects and high service quality offered by the country and the IRs actually brought in considerable revenue (from both gaming and non-gaming activities) to the industry and related local sectors. As depicted in Figure 5.2, the GGR of the industry crowned Singapore as the second largest casino gaming jurisdiction in Asia and the world from 2011 to 2013 (which was below Macao but slightly higher than Las Vegas Strip) after the soft opening of the two IRs in 2010 (RWS in January and MBS in April).

However, similar to other casino jurisdictions in the ESEA in which the targeted patrons are from mainland Chinese, the anti-corruption and anti-extravagant policies of the Chinese government between 2014 and 2016 also affected the

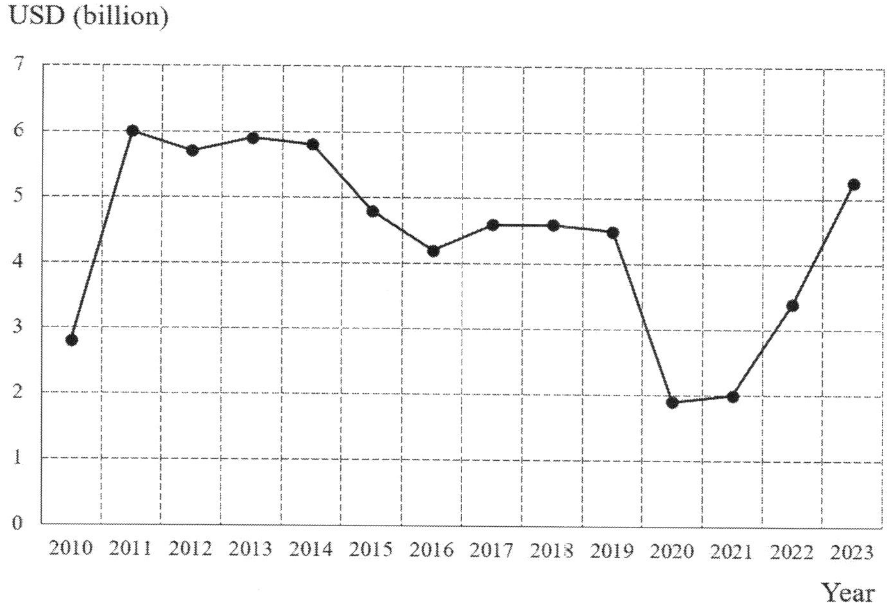

Figure 5.2 Changes in gross gaming revenue of Singapore from 2010 to 2023

Source: Author (data from 2010 to 2022 are derived from Nicholson (2023), data for 2023 from Gaming Intelligence (2024))

overall performance of casino tourism in Singapore. Nevertheless, since "VIP gaming – with the majority of punters made up of mainland Chinese high rollers – accounts for a smaller proportion of profit in Singapore" (Leong 2015), its downturn in this period was less severe than that in Macao (as compared to the slump depicted in Figure 4.1).[4] Owing to the changes in the Chinese government's policies to explicitly limit the spending of its residents on casino tourism outside the country, it was inevitable that the IRs in Singapore had also to adjust their business strategies by making further efforts to attract the mass population rather than the VIP segment in 2016 (Vasagar and Bland 2016). This adjustment allowed the industry to rebound without much issue before the outbreak of the COVID-19 pandemic in 2020.

The image of Singapore to the world clearly changed from an "unexciting" city to a "fun" city through the new attractions that were constructed and the promotions made by the two IRs. Thus, the number of international visitor arrivals to Singapore almost doubled between 2009 and 2019 (as depicted in Figure 5.3). Among the visitor arrivals, those from mainland China as a percentage of the total increased from 9.7% in 2009 to 19.0% in 2019.[5] Owing to the geographical location of Singapore, the primary source of its visitor arrivals was from the SEA, especially Indonesia, where casino gaming remains illegal.

Since China only accounted for less than 15% of the total number of visitor arrivals in the first half of the 2010s and non-gaming expenditures comprised a larger proportion of the total spending of tourists, it was observed that when visitor arrivals from mainland China and the GGR fell from 2013 to 2016 owing to

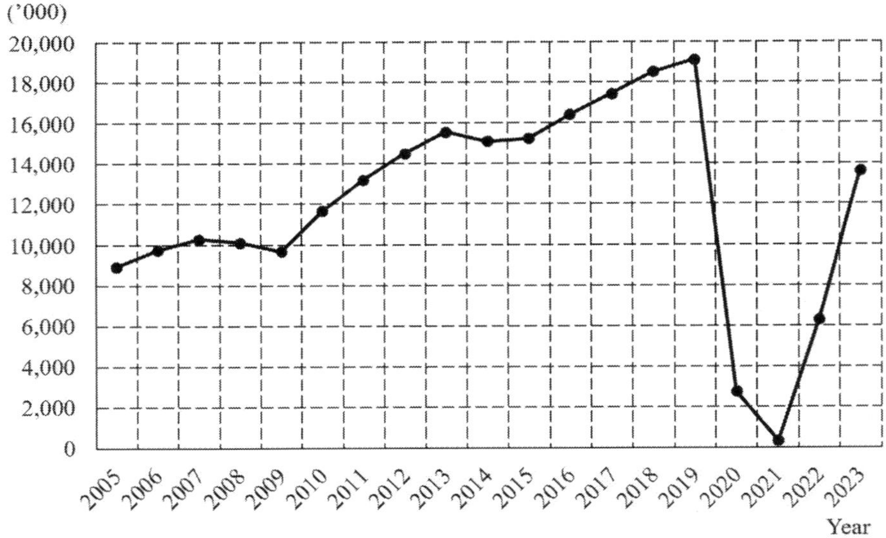

Figure 5.3 Number of international visitor arrivals to Singapore (2005–2023)

Source: Author (data from https://www.singstat.gov.sg/publications/reference/ebook/industry/tourism)

the aforementioned change in the Chinese government's policies, the total tourism expenditure remained consistent. In 2013, when the GGR was around USD5.9 billion, the total tourism expenditure was around USD24.4 billion. In 2016, the GGR and total tourism expenditure were around USD4.2 billion and USD24.0 billion, respectively (tourism expenditure actually rose to USD25.5 billion in 2014 before somewhat declining in 2015 and 2016).[6] Thus, the evidence shows that the practices of the IRs really met the objective of the government to "reinvent Singapore" (Ministry of Trade and Industry 2005a) as a vital city, especially its tourism industry.

5.3 Updated policies that align with the changing environment

Throughout the process of the development of casino tourism under the IR model, the performance of the industry and its interrelationship with the local economy and related feedback from society indeed provided crucial evidence for the Singapore government to review and update its policies with the expiration of the two IR licenses in 2017. In addition, changes (and potential changes) in the market environment, industrial layout and competition in the ESEA also forced the government to plan ahead strategically and with caution. Therefrom, the government decided to update the related policies and requirements for the ongoing operations of the industry. First, the government decided to retain the duopoly structure until 2030. This arrangement means that the two IR operators are committed to making new capital investments for the development of the industry (mainly in the expansion of non-gaming facilities) and increasing their tax payments to the government. In addition, casino entry levies and arrangements were also revised in an attempt to improve the effectiveness of the measures to reduce Singaporeans and permanent residents' interest in spending on casino gaming.

5.3.1 Retaining duopoly structure and ensuring ongoing investment

As a proactive authority, the Singapore government made it clear to the public in 2015 that the duopoly structure of the industry would be retained when the existing licenses expired in March 2017 (GGR 2015). Following this announcement, the process undertaken by the government to update the policies and requirements took longer than expected before the formal arrangements were released in April 2019.[7] Indeed, the government discussed with the two IRs and cautiously explored the future roadmap of casino tourism development in Singapore again for two years. In the process, the position of the government was made clear that since "both IRs contribute about 1 to 2 per cent of gross domestic product" (Heng 2019) to the country, they have to continuously support public interests if they see that the business value is worth their while to stay with the market in Singapore.

According to the formal documents released in 2019 (Ministry of Trade and Industry 2019; Ministry of Social and Family Development 2019), the two IRs promised to invest $S9 billion (around USD6.6 billion[8]) to expand their non-gaming facilities. These new investments are expected to bring in more tourists and create more job opportunities for the locals and business opportunities for the local business. Together with the agreement to increase casino entry levies and arrangements for local residents to access the casinos and increase the gaming tax, the two IRs were granted the exclusive rights to continue their operations until the end of 2030.

The $S9 billion in new investments represents "almost two-thirds of the IRs' initial investment of about $15 billion in 2006" (Lai and Tay 2019). Simultaneously, initial plans for these investments were outlined to the general public. These include the construction of a fourth iconic tower adjacent to the existing property (as presented in Figure 5.1) by the MBS, and the expansion of the existing Universal Studios Singapore by the RWS through the addition of two new theme parks (Minion Park and Super Nintendo) and other redevelopment and new construction projects in Sentosa (more details are found in Ministry of Trade and Industry (2019, 1)). Thus, it is evident that the attractions of casino tourism in Singapore will be enhanced, and thus its competitive edge in the region and the world, the real hospitality capacity and interplay with the local tourism sectors will also increase over time.

5.3.2 Raising taxes and casino entry levies

Due to their commitment to the new investments, the two IR operators also obtained government approval to expand their gaming facilities like the space of their gaming areas and gaming devices in the IRs mainly for entertaining high-end premium tourists. Nevertheless, as adverse social impacts may also potentially increase when the absolute scale of casino gaming is increased, gaming taxes were increased in a policy to allow the government to spend more on related social programs. To this, a two-tier progressive tax system was formulated and put into effect as of March 2022 to replace the previous flat rates on GGRs. As aforementioned, gaming taxes levied on the mass gaming and premium markets in 2010 were 15% and 5% of the GGRs. Under the new system, the first $*S*3.1 billion of GGR derived from the mass market is subjected to an 18% gaming tax, and any revenue that exceeds $*S*3.1 billion would incur a 22% gaming tax. As for the premium market, the gaming tax on the first $*S*2.4 billion of GGR is 8% and revenue that exceeds $*S*2.4 billion is subjected to a 12% tax (Ministry of Trade and Industry 2019, Table 1).

In view of the likely adverse social impacts from the expansion of the absolute scale of casino gaming and based on some related studies, the government also updated/tightened the policy immediately on August 1, 2019, to lower the potential spending of Singaporeans and permanent residents in the casinos. To this, the daily casino entry levy increased from $*S*100 *to* $*S*150, while the annual pass was increased from $*S*2000 *to* $*S*3000. In addition, the pre-payment and activation periods of the daily and annual entry levies were also tightened. In particular, local patrons "cannot purchase a 2nd Annual Levy or Daily Levy unless the 1st Annual Levy or Daily Levy is expiring within six hours" (Ministry of Social and Family Development 2019, 1). Again, despite whether tightening the policies would really reduce problem gambling issues, the government still makes ongoing efforts to update/improve the preventive measures to minimize the possible adverse impacts of this industry on society.

5.4 A new model for casino tourism?

Given the particular contextual settings in Singapore, it is undeniable that before casino tourism was initiated in Singapore, the government designed and carefully thought out a reasonable and realistic plan for its development. In the development process, the organization and practices of the IRs in Singapore actually established a new paradigm (i.e., the Singapore model) to better balance the potential benefits and costs from the development of casino tourism although the adverse social impacts like problem gambling and related crime associated with the operation of casino gaming are still inevitable. Based on the layout of this model, it is argued that "IRs are basically 'one-stop' entertainment centers and major mutations from 'gaming-centric' casinos. The potential economic benefits and social cost generated by IRs are radically different from gambling-centric casinos" (Leu and Ko 2012, 159).

5.4.1 Nuances of the Singapore model

In practice, a key attribute of the Singapore model is that the IR operators have the "freedom" to conduct their businesses and formulate their business routines only within the boundaries outlined and approved by the government at the outset. In other words, "[t]he government has played a leading role in shaping the character and running of the resorts, imposing conditions in an effort to ensure visitor attractions in conformity with official visions" (Henderson 2016, 203). To ensure the success and sustainability of this model, however, proactiveness and long-term vision of the government in establishing an effective and enforceable regulatory system which aligns with a set of well-specified public objectives and the contextual settings of the society concerned are needed.

In view of the social structure and stage of social and economic development in Singapore, the government clearly emphasized social safeguards when exploring the IR concept to receive economic benefits through the development of casino tourism. So, the aforementioned efforts (as described in the last section) of the Singapore government in constructing an effective regulatory system "to prevent their citizens from becoming problem gamblers" (Lee 2017, 17) are generally recognized by the international society. As a matter of fact, it is reported that the country's "prevalence of problem gambling reduced from 2.1% in 2005 to 0.2% in 2014" (ibid.).

On top of minimizing the adverse social impacts, Figure 5.2 also reveals that stability instead of rapid growth (which may be associated with greater volatility) in the casino gaming sector is the public preference for the industry's performance. For example, the growth of Singapore between 2011 and 2014, and 2017 and 2019 was relatively stable especially compared to the case of Macao as depicted in Figure 4.1, except for the restrictive policy imposed by the Chinese government in the mid-2010s and the unexpected outbreak of the COVID-19 pandemic in 2020, which reduced the GGR of the industry. Actually, the mechanism which maintained the stability of the business performance of the gaming sector could be largely attributed to the strict regulations that oversaw the business activities of the

IR operators and gaming agents in approaching premium players. Hereupon, notwithstanding the highly volatile nature of the VIP business segment which is easily to be affected by various external factors like the changes in macroeconomic environment and government policies in other countries, the gaming and non-gaming business performances of the IRs in Singapore are relatively stable with reasonable growth over time.

5.4.2 Implications of the Singapore model for other jurisdictions

Following the success of Singapore in developing its casino tourism, application of the IR concept is seen to be widely pursued by the other existing jurisdictions in the ESEA to revamp/rebrand the images and practices of their industries, as well as by potential jurisdictions to explore if and how they can reap the benefits from developing this industry with the least adverse impacts on their societies. For example, some recent studies indicate that the future development of casino tourism in South Korea should pay more attention to finding related capital investment by the industry and minimizing the adverse social impacts (Lee 2017). Besides, the process of the Japanese government in legalizing its IRs with casino gaming indeed exhibits a similar footprint as that of Singapore (Philander et al. 2017).[9]

Despite valuable experiences that could be learned from the practices of the IRs in Singapore and imitated by other jurisdictions in the ESEA, it is important to highlight that the direct reproduction of the Singapore model to other casino tourism destinations may not necessarily result in the same performance and level of success owing to the different contextual settings and the stage of social and economic development in the different jurisdictions. Indeed, this argument is similar to the experiences that simple replication of the Las Vegas-style mega casino resort model by other states like Atlantic City in New Jersey, the United States of America, may not guarantee their success in the long run. For optimal results with the Singapore model, the most essential lesson to be learned is the rationale and approach adopted by the government to effectively include this socially controversial industry into its economic society.

By referring to this model, the objectives of the development of casino tourism in the country (including the scale and scope of the businesses and their anticipated contributions to the local economy) have to be unambiguously identified at the beginning. Then, based on the structure of the society and its legal system, the approach used to regulate and supervise firms and related participants needs to be formulated realistically which definitely includes the design and enforcement of the related laws and regulations. From these, the IRs would then be established as the organizations to meet the objectives under a pre-determined set of rules. Finally, ongoing communication with the industry and periodical reviews and updates of the effectiveness of the regulations by the government are necessary to sustain the competitiveness of the industry and its real contribution to the economy. Actually, the evidence described in Section 4.4 also shows that recent revisions undertaken by the Macao government to revamp the industrial organization of its casino tourism also point to a similar rationale and approach as the Singapore model.

5.5 Summary

The path of development and frontiers of casino tourism in Singapore have fundamental differences in comparison to Macao as presented in Chapter 4. In Macao, casino gaming clearly has a dominant role in the tourism industry of the city and its economy. However, casino gaming in Singapore only represents a very small share of the aggregate economic income. In practice, the evidence presented for the process of development of casino tourism in Singapore shows that a highly and effectively regulated environment would be beneficial to not only the development of this socially controversial industry but also ensure its active interplay with the other local tourism sectors, thus maintaining the vitality of the tourism industry as a whole and its sustained contributions to the economy.

From a business standpoint, the prospects of this market are positive although the practices of IRs in Singapore are subject to strict regulations and high business costs to meet the required social responsibilities (especially responsible gaming and business integrity). To a large extent, consistent growth of business turnovers in normal periods of time could be expected not simply because the IRs have promoted Singapore as a "fun" city but because the integrity of the market has been instilled among international visitors. Besides, there is a high degree of diversity in the origins of visitors in the Singapore market owing to its location which is midway in between East and South Asia (particularly between China and India). This geographical factor also provides an advantage to Singapore in the development of its casino tourism.

Taking into consideration the particular SCPE forces and geographical factor which launch the particular environment for the actions taken by the Singapore government and the IR operators for the success of the Singapore model, it is evident that when other jurisdictions want to learn from this approach, finding a way to formulate the right model which fits into their respective contextual settings rather than simply constructing large-scale IRs in their countries is necessary to assure that the practices of the IR concept are helpful for the development of casino tourism.

Notes

1 "The Chinese community . . . account(s) for about 74 per cent of Singapore's population" (Soon et al. 2018, 9).
2 In 2004, USD1 = $S1.69.
3 These entry levies were raised in 2019 and related arrangements were also modified. Discussions are presented in Section 5.3.
4 As compared to the peaks in 2013, the GGR of Macao fell by 38% in 2016, while the GGR of Singapore fell by 29%. In the same period, the reported GGR of the Las Vegas Strip was increasing.
5 Measured with data from https://www.singstat.gov.sg/publications/reference/ebook/industry/tourism.
6 GGR data derived from Nicholson (2023) and tourism expenditure data derived from https://www.indexmundi.com/facts/singapore/indicator/ST.INT.XPND.CD.

7 Between 2017 and 2019 when the licenses had already expired and the IRs did not have a new agreement with the government, they were authorized to continue to operate under the existing terms.
8 In 2019, the average exchange rate of $S1 was around USD 0.7331.
9 The cases of South Korea and Japan will be examined in the following chapters.

References

Chia, Eng Seng. 2007. "Is the Integrated Resort a Good Gamble?" In *IEEE International Symposium on Technology and Society*, 1–7. IEEE. Available at https://ieeexplore.ieee.org/stamp/stamp.jsp?arnumber=4362212. Accessed July 19, 2020.

Chia, Joshua Yeong Jia. n.d. "Gambling in Singapore." Available at https://www.nlb.gov.sg/main/article-detail?cmsuuid=6860d586-4613-4172-ad48-c8a47efcf30b. Accessed June 12, 2022.

Eadington, William R. 1998. "Contributions of Casino-Style Gambling to Local Economies." *The Annals of the American Academy of Political and Social Science* 556 (1): 53–65.

Eadington, William R. and Eugene Martin Christiansen. 2007. "Tourist Destination Resorts, Market Structures, and Tax Environments for Casino Industries: An Examination of the Global Experience of Casino Resort Development." *Proceedings. Annual Conference on Taxation and Minutes of the Annual Meeting of the National Tax Association*, Vol. 100, 100th Annual Conference on Taxation (November 15–17, 2007): 225–235.

Gaming Intelligence. 2024. "Singapore Casino Revenue Tops S$5 Billion in 2023." (February 27, 2024). Available at https://www.gamingintelligence.com/finance/results/188846-singapore-casino-revenue-tops-s5-billion-in-2023/. Accessed March 12, 2024.

Getforme Singapore. 2004. "Social Safeguards for Integrated Resort with Casino Gaming." Available at http://www.getforme.com/previous2004/301204_socialsafeguardsintegratedresortwithcasinogaming.htm. Accessed July 24, 2022.

GGB (Global Gaming Business). 2010. "Singapore Sets VIP Rules." (February 2, 2010). Available at https://ggbmagazine.com/article/singapore-sets-vip-rules/. Accessed June 5, 2010.

GGRAsia. 2015. "Singapore Has No Plans for More Casino Licenses: Minister." (May 12, 2015). Available at https://www.ggrasia.com/spore-has-no-plans-for-more-casino-licences-minister/. Accessed June 20, 2015.

Henderson, Joan C. 2006. "Betting on Casino Tourism in Asia: Singapore's Integrated Resorts." *Tourism Review International* 10: 169–179.

Henderson, Joan C. 2012. "Developing and Regulating Casinos: The Case of Singapore." *Tourism and Hospitality Research* 12 (3): 139–146.

Henderson, Joan C. 2016. "Integrated Resort and Tourism: A Singapore Perspective." *Asia-Pacific Journal of Innovation in Hospitality and Tourism* 5 (2): 191–208.

Heng, Janice. 2019. "Diversified Economy Strengthened Singapore's Position in IR Talks: Chan Chun Sing." *The Business Times* (April 5, 2019). Available at https://www.business-times.com.sg/international/diversified-economy-strengthened-singapores-position-ir-talks-chan-chun-sing. Accessed June 14, 2019.

Lai, Linette and Tiffany Fumiko Tay. 2019. "Integrated Resorts to Build New World-class Attractions, Exclusivity Period Extended to 2030; Entry Levies to Go Up from Thursday." The Straits Times (April 4, 2019). Available at https://www.straitstimes.com/singapore/integrated-resorts-to-build-new-world-class-attractions-exclusivity-period-extended-to. Accessed June 14, 2019.

Lee, Terence. 2005. "Internet Control and Auto-regulation in Singapore." *Surveillance and Society* 3 (1): 74–95.

Lee, Young Gi. 2017. "The 'Singapore Model' in Gaming: Applications in South Korea." *UNLV Theses, Dissertations, Professional Papers and Capstones*, 3203. Available at http://dx.doi.org/10.34917/13444550. Accessed July 19, 2020.

Leong, Grace. 2015. "Singapore Casinos on a Roll Despite Fall in Chinese High Rollers." The Straits Times (May 22, 2015). Available at https://www.straitstimes.com/business/singapore-casinos-on-a-roll-despite-fall-in-chinese-high-rollers. Accessed June 18, 2015.

Leu, Gwo-Jiun Mike and Yiu Chung Ko. 2012. "A Policy Innovation for Social and Economic Development: The Case of Integrated Resorts in Singapore." *Social System Studies* 24 (March): 151–179.

Loke, Angela. 2004. "The Casino Debate: To Build or Not to Build." (November 26, 2004). Available at https://www.skyscrapercity.com/threads/the-casino-debate-to-build-or-not-to-build.154376/page-18. Accessed April 18, 2005.

Mei, Tan Tam. 2019. "Casino Patrons Could Receive Alerts on How Much Time and Money They Have Spent to Encourage Responsible Gambling." The Straits Times (August 29, 2019). Available at https://www.straitstimes.com/singapore/singapore-studying-responsible-gaming-measures-adopted-by-foreign-casino-regulators. Accessed December 16, 2019.

Ministry of Social and Family Development. 2019. "Annex A: Information on Social Safeguards." Available at https://www.mha.gov.sg/docs/default-source/media-room-doc/annex-a-media-factsheet.pdf. Accessed August 5, 2023.

Ministry of Trade and Industry. 2005a. "Statement by Prime Minister Lee Hsien Loong on Integrated Resort on Monday, 18 April 2005 at Parliament House." Available at https://www.nas.gov.sg/archivesonline/data/pdfdoc/2005041803.htm. Accessed July 24, 2022.

Ministry of Trade and Industry. 2005b. "Minister-in-Charge of Muslim Affairs Dr Yaacob Ibrahim's Speech on the Muslim Community's Concerns about Bambling." Available at https://www.mti.gov.sg/Resources/feature-articles/2012/Additional-Information-on-the-Integrated-Resorts. Accessed July 24, 2022.

Ministry of Trade and Industry. 2012. "RFPs for Developing Integrated Resorts in Singapore." (July 9, 2012). Available at https://www.mti.gov.sg/Resources/feature-articles/2012/RFPs-for-Developing-Integrated-Resorts-in-Singapore. Accessed July 24, 2022.

Ministry of Trade and Industry. 2019. "Integrated Resorts to Invest S$9 Billion in New World-Class Attractions and Experiences." (April 3, 2019). Available at https://www.mti.gov.sg/-/media/MTI/Newsroom/Press-Releases/2019/03/Press-release-on-IRs.pdf. Accessed June 17, 2023.

Nicholson, Matthew. 2023. "G2E Asia: Singapore Expected to Return to Pre-Covid Levels by 2024." (May 30, 2023). Available at https://www.gamblinginsider.com/news/21281/g2e-asia-singapore-expected-to-return-to-pre-covid-levels-by-2024. Accessed August 10, 2023.

Pearce, Gouglas G. 1978. "Form and Function in French Resorts." *Annals of Tourism Research* 5 (1): 142–156.

Philande, Kahlil, Brett Abarbanel, Bo Bernhard and Ray Cho. 2017. "Socio-Economic Impacts of Japanese Integrated Resorts." International Gaming Institute (IGI), University of Nevada, Las Vegas (UNLV). Available at https://dhhl80.hawaii.gov/wp-content/uploads/2021/01/Philander-Abarbanel-Bernhard-and-Cho_2017.pdf. Accessed July 19, 2020.

Song, Anthony. 2012. "CRA Awards Licences to Two Junket Operators." (March 22, 2012). Available at http://wealthadvisersg.blogspot.com/2012/03/cra-awards-licences-to-two-junket.html. Accessed April 7, 2012

Soon, Su-Chuin, Elvin Yifu Xing and Chee Kiong Tong. 2018. "Chinese Community and Culture in Singapore." In *The Singapore Ethnic Mosaic: Many Cultures, One People*. Edited by Mathew Mathews, 9–103. Hackensack, NJ: World Scientific Publishing Co.

Stanton, John and Colin Aislabie. 1992. "Up-market Integrated Resorts in Australia." *Annals of Tourism Research* 19 (3): 435–449.

Szep, Jason. 2005. "Singapore Gives Green Light to Two Casinos." (April 8, 2005). Available at http://www.boston.com/news/world/asia/articles/2005/04/18/singapore_gives_green_light_to_two_casinos/. Accessed April 18, 2005.

Tan, Sin Liang. 2011. "The Laws Governing Casinos in Singapore." Available at https://v1.lawgazette.com.sg/2011-01/feature2.htm. Accessed June 11, 2011.

Vasagar, Jeevan and Ben Bland. 2016. "Singapore's Casinos Shift Bets to Mass-Market Gamblers." *Financial Times* (June 21, 2016). Available at https://www.ft.com/content/18dd85c8-31ed-11e6-bda0-04585c31b153. Accessed July 4, 2016.

Winslow, Munidasa, Christopher Cheok and Mythily Subramaniam. 2015. "Gambling in Singapore: An Overview of History, Research, Treatment and Policy." *Addiction* 110 (9): 1383–1387.

Yap, Wai Ming. 2007. "Bringing Casinos to the Shores of Singapore." *Gaming Law Review* 11 (4): 440–448.

Yeo, Shiguang, Ching Chieh Low and Andy Chee Kheong Low. 2012. "A Study on the Impact of Entry Levies Implemented on Casinos in Singapore." Available at https://hdl.handle.net/10356/48152. Accessed June 22, 2023.

Additional readings

Bafoil, François. 2014. "State Liberalism and Market Socialism: A Comparison between Singapore and Vietnam." In *Emerging Capitalism in Central Europe and Southeast Asia: A Comparison of Political Economies*. Edited by Bafoil François, 65–86. New York, NY: Palgrave Macmillan.

GGB News. 2019. "Singapore Extends & Expands Two Casinos." (April 7, 2019). Available at https://ggbnews.com/article/singapore-extends-expands-two-casinos/. Accessed June 15, 2019.

Henderson, Joan C. 2015. "Destination Development and Transformation: 50 Years of Tourism After Independence in Singapore." *International Journal of Tourism Cities* 1 (4): 269–281.

Koh, Gillian. 2005. "Casino Debate: Laying Out All the Cards." *The Straits Times* (January 14, 2005), 14. Available at https://lkyspp.nus.edu.sg/docs/default-source/ips/pa_gillian_st140105_p27_casino-debate1.pdf. Accessed December 14, 2005.

Lee, Derrick, Abhishek Bhati and Laurie Murphy. 2013. "Singapore's Ability to Maintain Destination Competitiveness Through Integrated Resorts (IRs)." In *Proceedings of the International Conference on Managing the Asian Century*. Edited by Purnendu Mandal, 547–558. Available at https://link.springer.com/book/10.1007/978-981-4560-61-0. Accessed July 19, 2020.

Tulshyan, Ruchika. 2010. "Analysis: Why Singapore Went into the Casinos Industry." *The World* (May 15, 2010). Available at https://theworld.org/stories/2010-05-15/analysis-why-singapore-went-casinos-industry. Accessed July 24, 2022.

Yong, Kwek Ping and Kevin Wenhao Fu. 2006. "With the Two Integrated Resorts (IRS) in Singapore Due for Completion in the Next Few Years, How Should the IRS Align Their Strategies to Attract the Increasingly Affluent and Outbound Tourists and Businessmen from China?" International Conference on Gaming Industry and Public Welfare. Available at http://dx.doi.org/10.2139/ssrn.947523. Accessed December 17, 2006.

Zhang, Juan and Brenda S. A. Yeoh. 2017. "The State of Fun? Exclusive Casino Urbanism and Its Biopolitical Borders in Singapore." *Pacific Affairs* 90 (4): 701–723.

6 Different approach and performance

Malaysia and South Korea

As argued in Chapter 1, the frontiers of the development of casino tourism are largely facilitated by the SCPE forces. In the real world, governments adopt approaches at the outset that clearly differentiate the practices, since the social structure, cultural attributes, political interests and economic realities (including the structure of the economy and the stage of its development) could be quite different between countries/regions, and hence the performance of the casino gaming industry across different jurisdictions. While some major differences can be found by comparing the cases of Macao and Singapore, the development of casino tourism in other ESEA destinations also provides interesting and valuable evidence that enhances the current understanding of the frontiers of the development of this industry. For example, among the earliest jurisdictions that legitimized the operations of casino hotels/resorts (at the end of the 1960s and in the early 1970s), the respective contextual settings and particular roles of the governments in Malaysia and South Korea (hereinafter Korea) and their interactions with the industries unarguably determined the scale and scope of business for the development of this industry in their own country. However, their business practices and performances are different.

In this chapter, evidence derived from Malaysia and Korea is scrutinized to reveal the factors that shape the footprint and define the boundaries for the development of their casino tourism industry. To this, the reasons of the Malaysia government for approving a monopoly structure and establishing the requirements for the business operations of casino tourism in the country are analyzed in the first section. Given the particular social structure and public interests, the success of casino tourism in Malaysia is commonly attributed to the development of its casino gaming monopoly – the Genting. Then, its business practices and progress are elaborated in the second section. These include the strategies that the Genting uses to sustain its business in Malaysia and venture outward to the overseas markets, and the effects of the feedback on the overall performance of the industry.

The case of Korea is presented in the third and fourth sections. First, the decision of the South Korean government to legalize casino gaming in certain hotels exclusively for foreign visitors since the late 1960s and the progress of the industry are explored. Then, the forces and rationale that have led to the approval and opening of the landmark casino resort in Korea – Kangwon Land – in 2000 as the

DOI: 10.4324/9781003279884-6

only property (a monopoly) that is allowed to provide casino gaming services to the local Koreans in a particular region are examined. Besides, the performance of Kangwon Land and impacts of its operations on society and the long-term development of the industry are evaluated. To conclude, comments are provided on the prospects of casino tourism and potential changes in the frontiers of the development of this industry that reflect the experiences of these two jurisdictions.

Takeaways from this chapter:

1. The development of casino tourism in a hilltop region in Malaysia represents the development of tourism through the alignment of entrepreneurship with political interests.
2. The industrial structure and organization for the development of casino tourism in Malaysia are largely framed by the fact that it is a Muslim-dominated society, which is unarguably reflected in the political interests and legal system.
3. Genting Highlands effectively sustains its casino gaming monopoly and development in Malaysia through investment, thus accentuating the non-gaming attractions of the destination.
4. The venturing of Genting Highlands beyond Malaysia shows that internationalization is an effective strategy to diversify the nuances of casino tourism in the ESEA with less focus and reliance on casino gaming in an individual market.
5. The approval of foreigner-only casinos established in some of the hotels in the major tourism cities/regions of Korea attracts inbound international tourists and their spending.
6. The development of the largest casino resort in Korea, the Kangwon Land casino resort, is a particular objective of the Korean government to reurbanize/redevelop a declining regional economy in the Gangwon-do Province.
7. The approval of Kangwon Land to admit local Korean patrons is an important milestone to ensure its business turnovers, thus meeting public interest in contributing to the local economy.
8. An ineffective corporate governance structure and internal control system prevent real contributions from the operations of Kangwon Land to society.

6.1 Forces that have led to monopoly structure in Malaysia

Similar to other societies, gambling activities have long existed in Malaysia. Although research and publications about gambling in Malaysia are limited, “gambling was ‘brought’ to Malaysia by Chinese merchants in the 19th century” (Rathakrishnan and George 2021, 32). Nevertheless, most gambling businesses were illegal in the country except for horse racing, which was introduced “by the British during the 1800s” (ibid.), and lotteries, which were formally approved in 1952. Besides, a single casino gaming license was granted to a firm called Genting at the end of the 1960s. The particular social structure of Malaysia as a Muslim-dominated society is a primary concern of the Malaysia government in approving the legitimacy of gambling businesses. Owing to public concerns, the approach undertaken by Genting with its casino gaming monopoly to develop the gaming and non-gaming tourism facilities and operate these businesses in practice exhibits clear differences in comparison to other casino tourism jurisdictions.

6.1.1 Entrepreneurial efforts and reality

Following independence from the British government and the establishment of the Federation of Malaysia in 1957, the Malaysia government emphasized and incited economic development through industrialization (i.e., the manufacturing industries). Thus, the approval of a casino gaming license during the 1960s seemed questionable to the general public. This decision is especially critical in a society in which the population is dominated by Muslims, whose religion, Islam, prohibits gambling. In 1960, Muslims accounted for 50.04% of the total population in Malaysia versus the 16.34% in Singapore (Kettani 2010, 148). However, related documents show that the realization and success of casino tourism in the state of Pahang in Malaysia are actually attributed to the ambition and operations of an entrepreneur who obtained political consensus to initiate the development of this industry. Nevertheless, social feedback and changes in political interests blocked the expansion of this industry.

The "Genting story" is commonly cited as a unique case in the development of casino tourism in Malaysia (see, e.g., Firzanah (2021) and TravelisMalaysia.com. (n.d.)). In practice, the Genting story reflects the success of the founder (Lim Goh Tong) of Genting Highlands Berhad (now part of the international conglomerate Genting Berhad, or the Genting Group, which operates not just casino resorts in various countries but also plantations and energy businesses under the company names of Genting Plantations and Genting Energy), whose ambition and willpower were to innovatively develop (and transform) an attractive tourism destination on a hilltop region with impressive natural and scenic views. Geographically, this hilltop is located on the peak of Mount Ulu Kali (or Gunung Ulu Kali), which is 1,800 meters above sea level and about 58 kilometers from the capital of the country – Kuala Lumpur. Through effective deals with the government to invest in needed infrastructure like roadways to ensure that the destination could be efficiently accessed by local and international tourists, the proposal was approved and initiated in 1965.

To realize the project, Genting spent almost four years to first open up a road to the hilltop, which was completed in 1969, before construction of the first hotel commenced. Through these initial efforts, Genting received government approval to include casino gaming as a business segment on its property "to help accelerate the development of this remote area" (TravelisMalaysia.com. n.d.). Thus, the opening of the first Genting hotel, the Highlands Hotel (which was renamed Theme Park Hotel at the end of the 1990s), and Casino de Genting (Genting Casino) in 1971 clearly marked the dawn of casino tourism in Malaysia. To the development of the Genting Highlands thereafter, it is observed that while casino gaming generated considerable business turnovers for the firm, its renowned reputation in the country and overseas has been largely established through its continuous investment in non-gaming attractions, which include various featured theme parks and facilities.

6.1.2 Role of government and influence of particular social settings

Owing to the desire of the Malaysia government to explore other industries in the 1960s, such as tourism, to promote the modern development of the country, the proposal from Lim was accepted, and his ability to develop the project (between 1965 and 1969) without burdening the government, in turn, gained the trust and support of the government. For example, it is observed that the Malaysia government included "the principal objectives of tourism into the country's 2nd Malaysia Plan 1971–1975 (which) further emphasized the role of tourism in the economy" (Ministry of Tourism, Arts and Culture 2021). The tourism section of the 2nd Malaysia Plan specifies that the "rapid increase in the number of tourists coming into Malaysia will provide a strong incentive for the private sector to increase investment, . . . in specific tourist projects" (Economic Planning Unit n.d., 168). To this, the Genting Highlands in the state of Pahang was specified as one of the four regions in the 2nd Malaysia Plan to accelerate tourism development in the country.[1] So, Genting clearly had the opportunity to expand on an ongoing basis as planned.

During the expansion of the industry in the 1970s, it was evident that the government might have overlooked the possible downsides associated with the industry and criticisms from the Muslim community. Also, the Muslim population increased in the two decades that followed, from 50.04% of the total population in 1960 to 52.93% in 1980 and then 58.62% in 1990 (Kettani 2010, 148). Although gambling is prohibited in Islam, "there wasn't a law against Muslims frequenting the casino" (Badd 2023[2]) when the license was approved to Genting. As a matter of fact, contributions from the Muslim patrons and employees to the business of Genting Highlands triggered red flags in the community. Therefore, "people started calling for the government to protect the sanctity of Islam through laws" (ibid.) in the beginning of the 1980s. At the same time, as the government also expressed its willingness to defend the interests of the majority, a formal decision was finally made that stated, "Muslims cannot enter the Genting casino" (ibid.), effective from June 8, 1983. To ensure the credibility and effectiveness of this decision, enforcement was carried out "by religious inspectors from the Pahang Religious Department along with Genting's treasury and management" (ibid.). Although it is still possible to identify cases where some Muslims occasionally venture into the casino, the law and its enforcement at least conveyed that the government was proactive in responding to the demands of society at large.

However, the persistence of the majority who wish to protect their community from the negativities of gambling means that the federal and local governments in Malaysia have to maintain a steadfast position over time to reject any ideas or proposals to issue a second license for casino gaming in other tourism destinations in the country (see, e.g., Malaysiakini (2003) and Phil (2014)).[3] Besides, in considering the many voices in society, especially the Muslim community, it is also clear that the government holds a similar position when dealing with the changes in the business scale of Genting Highlands in casino gaming. In other words, it is reasonable to anticipate that the capacity for Genting to expand its casino gaming business in Malaysia will be limited. For example, Genting Highlands reorganized its gaming venue without increasing more casino space (see, e.g., New Straits Times (2016)).

6.2 Business practices and dynamics of the industry

Since the development of casino tourism on Mount Ulu Kali in Malaysia has been largely initiated by an entrepreneur at the outset, the practices and progress of this industry in Malaysia will continue to transpire as long as legally permissible. To a large extent, evidence shows that Genting has responded in a positive and effective manner to the feedback from the community and government policies which contributes to sustaining its business development. In addition, the success of Genting actually explicitly contributes to the dynamics of casino tourism development in the ESEA although the business scale of casino gaming in Malaysia is limited by social interests. On top of expanding its business in the United Kingdom (UK) and the United States of America (USA), Genting also successfully contributed to the development of this industry in Singapore (as discussed in Chapter 4) and the Philippines (which will be discussed in Chapter 7). Although its attempts to enter the market in Macao in both 2002 and 2023 were not successful, its efforts to venture out beyond Malaysia with its relatively competitive "Genting" brand have undoubtedly generated the ongoing dynamics of this industry.

6.2.1 Positive responses to casino monopoly and its progress

Despite the crucial legal change in 1983 which prohibited Muslims from entering the casino, Genting continued to thrive and advanced the casino gaming industry in Malaysia. Between the 1980s and 2010s, Genting opened six more hotels and new resort facilities, which transformed the remote hilltop region into a comprehensive tourism destination with unique features. As depicted in Figure 6.1, Genting Highlands has unarguably and truly developed into a destination for visitors to spend multi-day vacations, while casino gaming is a complementary service for their enjoyment. Other than new hotels, visitors can visit theme parks and embark on cable car systems (called the Swana Skyway and Genting Skyway) to accentuate their experience during their visit.

As compared to tourism development in other regions with location advantages such as Kuala Lumpur (prime position on the west of the Malaysian peninsula) and Penang (located on the northwest coast of the Malaysian peninsula with rich natural, historical and cultural tourism resources), the unique positioning of Genting Highlands has unarguably allowed the company to gain reasonable growth and market share, although there was only a small flow of tourists in absolute number in its early stages. For example, Oppermann's (1992, 488 – 489) survey in 1990 shows that Genting Highlands was ranked as the third most welcomed overnight destination by international tourists when they visited Malaysia. Muhibudin et al. (2010, 206) find that following the ongoing expansion in non-gaming attractions/facilities, "Genting Highland (became) one of the most popular tourist destinations

Figure 6.1 The Genting Highlands

Source: https://www.viator.com/en-IN/tours/Kuala-Lumpur/Genting-Highlands-Day-Tour-from-Kuala-Lumpur/d335-5772KULTO6

Table 6.1 Share of gaming in total business revenue of Genting Highlands

Year	*Net gaming (as percentage of total leisure and hospitality revenue)*
2017	87.6%
2018	84.5%
2019	77.5%
2020	82.5%
2021	81.4%
2022	78.1%

Source: Raw data derived from Genting Malaysia (Annual Report, various years).

Release of breakdowns of the company's "net gaming wins"[4] commenced in 2017.

in Malaysia between 2001 until 2006. As the 'City of Entertainment', it provides theme parks and other types of entertainment for the whole family besides the main business of casinos."

To realize investment for the development of non-gaming facilities, it is evident that the business practices of casino gaming provide a considerable source of funds for the company. As shown in Table 6.1, revenue derived from casino gaming accounted for over three-quarters of the total revenue of Genting Highlands from 2017 to 2022. However, it is worth noting from the data presented in

Table 6.1 that the business composition of Genting Highlands is gradually diversifying, although society may demand that the company does so at a faster pace. Other than in 2020 and 2021, when tourism activities in the world markets were hit hard by the COVID-19 pandemic so that non-gaming spending slumped, the share of non-gaming revenue to the total leisure and hospitality revenue of Genting Highlands increased by almost 10% from 12.4% in 2017 to 21.9% in 2022. To this, the dynamics between gaming and non-gaming development in the development of casino tourism are stabilized (although adverse social impacts may still be there).

6.2.2 *Successful brand venturing beyond Malaysia*

Given the aforementioned limits of the expansion of casino tourism in Malaysia and the entrepreneurial spirit and ambitions of the founder of Genting, the leaders of Genting moved actively and pragmatically to extend their successful brand ("Resorts World Genting") to the world markets, including jurisdictions in the ESEA. Other than Resorts World Birmingham and other individual casino properties in various cities in the UK (e.g., London, Liverpool, Manchester, Birmingham), Resorts World New York City in the USA and Resorts World Bimini in the Bahamas, the opening of Resorts World Sentosa in Singapore in 2010 and Resorts World Manila in 2009 (which has been subsequently renamed as Newport World Resorts) in the Philippines not only represents the success of the company in their international expansion but also reflects the positive market potential of casino tourism for further development in the ESEA.

The success of venturing out beyond its home base to the international markets means that the leisure and hospitality businesses of Genting actually diversified more, and the risk to its overall business performance is reduced, even if the government policies in any of the venture countries change, particularly for casino gaming. First, it is shown in Table 6.2 that the reliance of Genting Berhad on casino gaming as a whole in the world markets is reducing, which shows a similar trend as Genting Highlands in Malaysia (see Table 6.1). Besides, a comparison of the

Table 6.2 Share of gaming in total leisure and hospitality revenue of Genting Berhad

Year	*Net gaming (as percentage of total leisure and hospitality revenue)*
2017	74.6%
2018	73.3%
2019	70.9%
2020	76.0%
2021	71.7%
2022	64.4%

Source: Raw data derived from Genting Berhad (Annual Report, various years).

Breakdowns of "net gaming wins" of company were released starting in 2017.

data in Table 6.2 with those in Table 6.1 shows that the casino tourism operations of Genting Berhad as a whole are less reliant on gaming than Genting Highlands which is only in Malaysia (all values in Table 6.2 are lower than those in Table 6.1).

Indeed, Genting has clearly expanded as a multinational corporation (MNC) in the global leisure industry (and casino tourism). In terms of total revenue derived by Genting Highlands (i.e., the leisure and hospitality businesses of Malaysia), its share of the corresponding business segments of Genting Berhad as a whole in the world markets decreased from 92.0% in 2005 to 37.7% in 2010, and further to 29.1% in 2022.[5] In other words, if government policies for casino gaming are tightened in an individual market, the adverse impact on the overall operations of the casino tourism business of Genting could be reduced. Last but not least, Genting as an MNC in casino tourism also has the ability to promote and further diversify the sources of patrons in some of the major destinations.

6.3 Casino tourism for foreign visitors of South Korea

As indicated in Table 2.1, South Korea is the first country in the ESEA which opened its first legalized modern casino hotel in 1967. By law, however, casinos in hotels are only allowed to receive foreign visitors so these properties are usually called “foreigner-only casinos” (or simply “foreigner casinos”) in Korea. Local Koreans are only permitted in one casino, the Kangwon Land Casino in Kangwon Land Resort, which opened in 2000 (forces that have led to this change and its contribution to the growth and development of casino tourism in Korea will be discussed in the next section). Basically, foreigner casinos are located in tourism cities and regions like Incheon, Seoul, Busan and Jeju Island.

While the decision of the government to legalize casino gaming for the enjoyment of foreign visitors aims to facilitate the development of international tourism in related cities/regions, the prohibition of Koreans in casinos is primarily based on the concerns of the possible adverse social impacts to the local community, especially the risk of gambling addiction by locals. Besides, although casino gaming is legalized as a business segment for tourism development, their operations were supervised strictly and directly by the Police Department until 1994 (Cho 2002, 188). In 1994, “the authority for casino gambling was transferred from the Police Department to the Department of Tourism at the Ministry of Culture and Sports in order to promote it as part of the tourism industry” (Lee and Kwon 1997, 53). This change indeed provided a basis for a later decision of the government to approve a new casino resort (Kangwon Land) which is accessible to both foreign visitors and local Koreans.

6.3.1 Casino gaming as an element for tourism development

When Korea started state-led industrialization and the modern development of its economy in the 1960s, the country was highly reliant on support from the USA, and traded with the USA and Japan. Thus, the inflow of visitors from these two and other countries subsequently increased. Besides, the government also introduced policies to attract international inbound tourists “as a strategic tool to obtain foreign currency” (Shin 2010, 527) even though Korea was being positioned as an export-oriented economy which focused on exports of primary and manufacturing commodities (see, e.g., Kim 1971). In line with public interests in fostering the development of international tourism, casino gaming was legalized which brought about the opening of the first casino hotel in Korea, the Olympos Hotel in 1967 in Incheon (known as the Paradise Casino Incheon today), which has rich tourism resources and located in the west coast. The hotel borders Seoul, the capital of Korea. As a form of convenient entertainment service provided to tourists when they are staying at the hotel, casino gaming has indeed increased its business turnover.

Following the favorable experience and manageable operations of the first casino hotel, a second casino, the Paradise Walkerhill Casino, was then approved in 1968 and located in the Sheraton Walkerhill Hotel in Seoul. Thereafter, three more casino hotels were approved in the 1970s, two in the 1980s, six in the 1990s and three in the 2000s (Cho 2002, 188, Table 1; Kim and Kang 2018, 4, Table 1). Recently, a casino resort was also approved in 2016[6] (Kim 2016). Of the 17 foreigner casinos that have been approved as of 2023, eight are located in hotels in a well-known tourism destination called Jeju Island situated 82.8 kilometers from the south of the Korean Peninsula which boasts beautiful scenery and natural resources. To a large extent, the practices of this business segment unarguably increase total tourism spending, especially that of the Chinese visitors although the primary motivations of many tourists who visit Jeju Island may not necessarily be casino gaming.[7]

Overall, foreigner casinos are approved as part of the entertainment facilities in various 4-/5-star hotels which are not large in absolute scale and may not offer a wide range of entertainment services as compared to casino resorts or IRs. Yet, it is commonly observed that the additional income derived from casino gaming actually creates a better financial position for these hotels so that they spend more on the exterior and interior designs of their properties which enhances their aesthetics, offer entertainment events like short shows, and provide higher quality services to accentuate the experiences of the guests during their stay. Besides, the legal right of the casino hotels to organize various gaming activities like baccarat tournaments for overseas tourists not only is helpful for increasing the occupancy rates of their accommodation and catering facilities, but also offers business opportunities for some of the neighboring local tourism-related firms.

6.3.2 *Performance and progress of foreigner casinos*

By incorporating casino gaming in their properties, foreigner casino hotels show in general their vitality and positive contributions to the development of tourism cities/regions. For example, the data reviewed in Kim and Kang (2015, 5–6) reveal that the compound annual growth rate (CAGR) of the total percentage of foreign tourists who visited Korea in the 20 year period between 1993 and 2013 is 6.70% versus the 7.39% who visited foreigner casinos. On the other hand, casino gaming revenue[8] derived by these casinos increased from *USD*173.176 million in 1993 to *USD*1.250 billion in 2013, which is a higher CAGR of 10.39% as compared to the increases in visitor arrivals to the country and casino guests respectively in this period. As so, it is evident that the foreign casinos were becoming more popular to international tourists in the said period. To a large extent, a major reason for this trend could be attributed to the increasing gaming expenditures of the Chinese visitors.

Despite the periodical fluctuations in the business performance of the foreigner casinos (see, e.g., Kim and Kang (2015, 6) for the period between 1993 and 2013 and Table 6.3 for period between 2013 and 2022) owing to external factors like changes in the economic environment of neighboring/world economies and policy restrictions imposed by the Chinese government on the its outbound tourist spending on casino gaming, as well as the COVID-19 pandemic in 2020, casino tourism in Korea shows a trend of long-term growth. For example, although the reported gaming revenue of the foreigner casinos between 2015 and 2017 fell below 2014, which is similar to other casino jurisdictions in the ESEA when the Chinese government imposed restrictive policies in that period of time, their business turnovers bounced back in 2018 and 2019 which were even higher than 2014 (see Table 6.3). Simultaneously, the number of foreign visitors the foreigner casinos in 2019 was also higher than that in 2014. Thus, it is reasonable to infer that the spending of foreign visitors on non-gaming tourism activities also increased over time.

In considering the practices of foreigner casinos and the path of development of casino tourism in the ESEA (especially the development of IRs), it is evident that one of the deficiencies of this industry in Korea is the lack of comprehensive non-gaming facilities which could be provided by the casino hotels within their properties. To this, the Korean government gave approval in 2016 to a leading casino resort operator in the USA, Mohegan Sun, to develop a casino resort under the concept/layout of an IR in Incheon. This decision definitely turned a new page in casino tourism in Korea (see, e.g., Kim (2016) and Wilhelm (2023)). Accordingly,

Table 6.3 Changes in gaming revenue in foreigner casinos, 2013–2022

Year	*KRW*[†] *(trillion) (a)*	*USD (billion) (a)/(b)*	*Annual average exchange rate of KRW per USD (b)*
2013	1.3685	12.498.02	1,094.97
2014	1.3772	13.074.75	1,053.33
2015	1.2433	10.986.52	1,131.66
2016	1.2757	10.990.80	1,160.70
2017	1.2072	10.242.79	1,178.59
2018	1.6253	14.768.12	1,100.55
2019	1.4489	12.433.83	1,165.29
2020	0.5978	0.506.95	1,179.20
2021	0.4050	0.353.69	1,145.07
2022	0.7145	0.552.95	1,292.16

Source: Gaming revenue in KRW, https://www.ngcc.go.kr/eng/stats/scale.do Exchange rate: https://www.exchangerates.org.uk/USD-KRW-spot-exchange-rates-history-2021.html

[†] KRW: unit of Korean currency.

the soft opening of INSPIRE Entertainment Resort (IER) at the end of 2023 and its casino in the beginning of 2024 (see, e.g., AGBrief Editorial (2023)) actually marks a new milestone for casino tourism development in Korea.

Although the casino gaming in the IER is also restricted to foreign visitors, its overall layout as an IR instead of a traditional/individual 4-/5-star hotel is unarguably more like a comprehensive non-gaming facility, and attractions like an arena, performances and Meetings, Incentives, Conferences and Exhibitions (MICE) to not just foreign individual and business visitors, but also local Koreans. From the academic and public policy standpoints, the practices of the IER may provide a valuable opportunity to explore the future development of casino tourism in Korea under a new blueprint of its foreigner casinos.

6.4 Casino monopoly for local Koreans

A comprehensive review of the literature on the development of casino gaming and tourism in Korea reveals that a large number of these studies were conducted and published in the 2000s with research interest related to the development and practices of the Kangwon Land casino resort which was launched in 2000 (see, e.g., Back and Lee 2005; Byun et al. 2012; Cho 2002; Lee 2010a, 2010b, 2017; Lim and Hong 2016). Unlike foreigner casinos, the Kangwon Land Casino is part of a comprehensive resort (the Kangwon Land Resort) in which the casino is approved to receive not only foreign visitors but also local Koreans. From the perspective of local Koreans, the Kangwon Land Casino is unarguably a casino monopoly as this is the only venue where they can legally partake in casino games in the country. Besides, as a destination by itself, the operations of Kangwon Land indeed contribute to the inclusive development of casino tourism in a particular region.

As compared to the establishment of foreigner casinos in the tourism cities/regions, the underlying force that has led to the development of the Kangwon Land casino resort is indeed unique in the urban redevelopment decision made by the Korean government at the end of the 1990s. To this, the economic interests of the government carried out through the operations of Kangwon Land clearly influence its business organization and practices. Nevertheless, the role taken by the government was not effective in the initial stage so the real contribution of Kangwon Land to the local society and economy was controversial, which included social issues associated with the increasing gambling problems of the local Koreans.

6.4.1 Rationale and strategic issues with the development of Kangwon Land

Geographically, Kangwon Land is located in the Jeongseon-gun county of Gangwon-do Province in the northwest of Korea,[9] where a large part of the region is “mountainous, making it difficult to access, but the terrain offers tourism opportunities for winter sports” (Wu 2017, 61), like skiing. The economy of Gangwon-do Province relied heavily on the coal mining industry before the 1990s. However, owing to the nationwide policy to eliminate coal mining and hence close down most of the coal mines in the country in the first half of the 1990s (Cho 2002, 189 and Table 3), the economy of Gangwon-do Province suffered a severe downturn and faced widespread unemployment. To reinstate the economy, the government promulgated a "Special Act on the Assistance to the Development of Abandoned Mine Areas" (Lim and Hong 2016, 5058) in 1998, which approved the construction of a destination resort, Kangwon Land in Jeongseon. In view of the geographical settings of the region, skiing and golfing are established as the primary entertainment facilities (attractions) for tourists.

Yet, transportation inconvenience to Kangwon Land which is located in a mountainous region would have reduced its attraction to both foreign and local visitors. Thus, part of the development of Kangwon Land was to ideally reurbanize the region and revitalize the local economy; otherwise, increasing local employment opportunities and government tax revenue would hardly be feasible. To increase the likelihood of the success of Kangwon Land, the government approved a crucial strategy, which was the establishment of a casino in Kangwon Land as the only venue in the country to offer casino gaming to local Koreans. Through this strategy, the advocates deemed that it would also be helpful to attract Korean gamblers who would "spend their money within Korea, rather than taking it abroad to gambling paradises like Macao or Las Vegas" (Zoll n.d.).

As a government-initiated project to redevelop/revitalize the declining economy of the Gangwon-do Province and the political interest to gain control over the economic benefits from the operations of Kangwon Land, the capital stock of Kangwon Land "is 51 percent owned by national and local government agencies, including the Coal Industry Promotion Board, which holds 36 percent" (Las Vegas Sun 2001). Nevertheless, public ownership of a casino gaming firm may be at risk of ensuring effective supervision. Actually, evidence shows that even though the government-appointed management teams of Kangwon Land are able to derive the monetary returns as desired, business integrity and social acceptance are in question (or even being criticized).

For example, issues associated with increasing gambling addiction problems of local Koreans have not been well addressed by Kangwon Land. Besides, the top management corruption and detected money-laundering issues associated with its practices have been found and reported over time (see, e.g., Lim and Hong (2016) and Malenovic (2023)). Even though the government attempted to improve its leadership in 2014 (see, e.g., Lim and Hong (2016)), the interests of the different government agents in the operations of Kangwon Land might continue to prevent any real improvement in corporate governance, especially the controversial and adverse social impacts from its business practices.

6.4.2 Business performance of Kangwon Land and its impacts

As a casino monopoly that caters to local Koreans, the Kangwon Land Casino immediately became the largest casino in Korea in terms of the reported casino gaming revenue (GGY, see Footnote 8). In 2006, the reported GGY of Kangwon Land represented 61.3%[10] of the total casino gaming revenue in Korea. In other words, only 38.7% of the GGY of Korea in 2006 was obtained from the

16 foreigner casinos. Owing to the rapid growth in the gaming expenditures of Chinese visitors in the foreigner casinos in the second half of the 2000s and beginning of the 2010s (i.e., before the Chinese government imposed its anti-corruption and anti-extravagant policies in 2014), the GGY of Kangwon Land Casino indeed increased from USD0.84 billion in 2006 to USD1.17 billion in 2013, although its share of the total GGY in the country fell to 48.3% in 2013.[11] Under the denomination of the Korean won, the reported GGY by Kangwon Land continued to grow between 2014 and 2017, while the foreigner casinos recorded a decline in their GGY in this period of time.[12]

The success of Kangwon Land and its contribution to the economy can be generally confirmed by using monetary terms at the macroeconomic level. For example, two statistical analyses (Lee and Lee 2009; Lee 2010a)[13] released in 2009 and 2010, with data provided by the Bank of Korea in 2007 and 2009, respectively, have basically consistent results and show that the operations of the Kangwon Land casino resort contributed directly and indirectly to the overall economic performance of the Gangwon-do Province. These include not only income derived directly by the Kangwon Land casino resort and hence various job opportunities offered to the locals but also contributions to government taxes and business opportunities to firms within and outside the casino resort and in the neighboring areas. Specifically, it is estimated that the "casino revenues of *US*$ 1,717 million in 2007 generated $4,042 million of output, $716 million of personal income, $2,200 million of value added, and $613 million of indirect tax throughout direct, indirect, and induced impacts. 26,922 full-time jobs were also created from casino revenues throughout ripple effects" (Lee and Lee 2009, 63).

While it produced economic benefits, however, the operations of Kangwon Land unarguably resulted in political conflicts and adverse social impacts. As aforementioned, corruption and money-laundering issues have been eroding the business integrity of Kangwon Land over time. In addition, as a casino monopoly for the local Koreans, their excessive gambling in the 2000s and beginning of 2010s caught the attention of the general public (see, e.g., GGRAsia (2016); Kim (2012); and Lee (2017)). As noted by Lee (2017, 21), "the rate of problem gamblers in South Korea was 6.1% in 2012, which was more than two times higher than the average rate of other foreign countries," like the UK and Australia.

To resolve these issues, effective restructuring of the leadership of Kangwon Land became necessary. Thus, the government appointed a new president in 2014 (see, e.g., Lim and Hong (2016)) in the attempt to improve the management of the largest casino resort in Korea. From this, some improvements were made, like the tightening of "its customer entry rules as a measure to prevent gambling addiction" (GGRAsia 2016), which included raising the entry levy for local Koreans and restricting the number of days within a month that a local patron could enter the casino. However, evidence also shows that enforcement of the related measures might be insufficient. For example, as compared to the practices of the "Singapore model," it is argued that a major difference in "South Korea's gaming regulation system was the level of social-safeguards for their domestic people" (Lee 2017, 24).

Besides, in 2023 Kangwon Land "has been imposed an overall fine of *KRW* 3.23 billion, which is approximately *US*$2.39 million, for 182 cases of violations of anti-money laundering (AML) laws" (Malenovic 2023). Accordingly, other than the adverse impact on problem gambling, corporate governance and internal control systems like the detection of money-laundering issues that are associated with the business practices of Kangwon Land have to be further improved. In a dynamic and an increasing competitive environment of the ESEA, drawbacks associated with casino operations of Kangwon Land can generate negative feedback effects on its attractions to foreign visitors and its non-gaming business performance as well. In addition, since the Chinese government has been increasingly concerned about the spending of the mainland Chinese in overseas casinos, unfavorable reports of the operations of Kangwon Land may also affect the image of foreigner casinos in Korea, thus limiting the development of the industry as a whole in the long run.

6.5 Summary

With reference to the evidence and discussions on the development of casino tourism in Malaysia and South Korea respectively, it is clear that while the principles (i.e., the SCPE forces) applied to show that the frontiers of the development of this industry are the same across various destinations, their practices are quite different. Actually, the different contextual settings between Malaysia and Korea are responsible for the differences in the path of development and the frontiers of development between these two jurisdictions. For example, the underlying forces which led to the establishment of Genting Highlands in Malaysia and Kangwon Land in Korea are clearly different. Besides, under different social structures, political interests and legal systems, the business practices and performances of casino tourism in these two countries actually reflect the attributes of their respective societies and changes over time.

Specifically, throughout the process of more than half a century of development, it is evident that the frontiers of the development of casino tourism under a monopoly structure in Malaysia are mainly framed by the ambition and operations of an entrepreneur and interactions of the leading team with the government in a society dominated by Muslims. On the other hand, the frontiers of the development of casino tourism in Korea advanced from the general establishment of foreigner casinos in major tourism cities/regions to the approval of a casino monopoly in a remote area which is authorized to admit local Koreans. Actually, this decision represents the public interest in facilitating the reurbanization and redevelopment of a region, thus saving/revitalizing its economy. Thus, the frontiers for the development of casino tourism (including the layout of non-gaming facilities) in the region are outlined accordingly.

In Malaysia, the social and cultural attributes (dictated by the Islamic faith) and their influences on the political interests and enforcement of related laws and regulations do not provide much room for the expansion of its casino tourism, especially the scale of casino gaming. In contrast, although social concerns in

Korea regarding the adverse impacts of casino gaming were restricted to the operation of small-scale foreigner casinos in major tourism cities/regions before 2000, approval and opening of a large-scale casino resort in a remote area that aimed to reap economic benefits by admitting local Koreans into its casino unarguably led to unfavorable social and political impacts. Therefore, a more effective regulatory system and its enforcement to supervise the business practices of Kangwon Land are necessary to sustain the development of this industry in Korea.

Despite the meltdown of the industry during the COVID-19 pandemic between 2020 and 2021, rebound of the markets in 2022 indeed conveyed favorable messages for further development of casino tourism. For example, continuous debates in Malaysia about whether it is reasonable and feasible to approve a second casino license in Sarawak and the opening of the IER in South Korea at the end of 2023 actually show that the markets still have positive expectations for the continuous growth of casino tourism in the ESEA. Nevertheless, owing to the increasing competition in this region and the changing attributes of the "China factor," which include the tightening of policies by the Chinese government to restrict the spending of the mainland Chinese on casino gaming outside China and changes in the gaming behavior of the new generations of Chinese, any crucial decision made by the governments of these two countries to expand the scale of this industry definitely requires careful and proactive investigation.

Notes

1 The other three regions indicated in the plan include Perak (Maxwell Hill and Pangkor Island), Selangor (Maimaland in Kuala Lumpur) and Johor (Skudia).
2 The facts and arguments presented in Badd (2023) are supported by a number of local newspaper clippings in chronological order.
3 At the time when this chapter was written in 2023, a new idea was being explored whether a second casino license could be approved in Sarawak, a state located in another island called Borneo (located to the east of the Malaysia peninsula), which is shared by Malaysia, Indonesia and Brunei. The idea is being hotly debated in the communities (see, for example, Annuar (2023)).
4 According to the company's annual report, net gaming wins represent the revenue derived from gaming operations which is the "net of complimentary goods and services provided to customers" (Genting Malaysia 2022, 127). In other words, this amount reflects the net gaming revenue instead of gross gaming revenue.
5 Calculated with data derived from the annual reports of the Genting Berhad and Genting Malaysia respectively.
6 This casino resort was opened at the end of 2023, and some discussions about this property are presented in the last part of this section.
7 Actually, the spending of Chinese high rollers in casino gaming increased rapidly in the second half of the 2000s and beginning of the 2010s.
8 In South Korea, data released on the official website (https://www.ngcc.go.kr/eng/stats/scale.do) for casino gaming revenue is termed as "gross gambling yield" (GGY), which refers to "the real expenditure from customers."
9 Kangwon Land is 150 kilometers (around three hours of driving) to the east of Seoul.
10 Calculated with data in Table 1 in Lee et al. (2018).
11 On average, $USD1 = KRW954.88$ in 2006, $USD1 = KRW1,094.97$ in 2013.

12 With reference to the data provided on the website of the Korean National Gambling Control Commission (https://www.ngcc.go.kr/eng/stats/scale.do).

13 The statistical method used in these studies is called the input-output analysis, which allows researchers to estimate direct, indirect and induced impacts (i.e., total outputs) generated from the spending (inputs) of firms in an economic sector. For example, inputs of the casino sector for its operations will bring about direct income to employees who work in the casinos. From this, accommodation and entertainment services provided to gamblers within the casino hotels/resorts and other local tourism sectors will lead indirectly to additional employment opportunities (hence higher income) in the economy. The spending of the employed groups will further motivate other businesses in the economy. Thus, the total income output (income) generated for the economy as a whole will be usually larger than the initial spending (inputs) by the casino gaming sector.

References

AGBrief Editorial. 2023. "INSPIRE Entertainment Resort in Incheon Is Scheduled for Soft Opening on November 30th, Casino in 1Q24." *Asia Gaming Brief* (November 24, 2023). Available at https://agbrief.com/news/south-korea/24/11/2023/inspire-resort-in-incheon-is-scheduled-for-soft-opening-on-november-30th-casino-in-1q24/. Accessed November 29, 2023.

Annuar, Azril. 2023. "Is a Casino in Sarawak a Good Bet?" *FMT* (May 13, 2023). Available at https://www.freemalaysiatoday.com/category/nation/2023/05/13/is-a-casino-in-sarawak-a-good-bet/. Accessed July 10, 2023.

Back, Ki-Joon and Choong-Ki Lee. 2005. "Residents' Perceptions of Casino Development in Korea: The Kangwon Land Casino Case." *UNLV Gaming Research &* Review Journal 9 (2): 45–53.

Badd. 2023. "Starting 8 June 1983, Malay Muslims Were Banned from Entering Genting's Casino." (June 9, 2023). Available at https://cilisos.my/starting-8-june-1983-malay-muslims-were-banned-from-entering-gentings-casino/. Accessed October 21, 2023.

Byun, Woo-Hee, Bon-Ki Koo and Timothy Jeonglyeol Lee. 2012. "Casino Development and Visitor Satisfaction: A Case of Korea." *Tourism Analysis* 17 (5): 659–671.

Cho, Minho. 2002. "Tourism Redevelopment Strategy: The Case of the Kangwon Land Resort Casino." *Anatolia* 13 (2): 185–197.

Economic Planning Unit, Prime Minister's Department. n.d. "Second Malaysia Plan, 1971–1975." Available at https://policy.asiapacificenergy.org/node/1273. Accessed October 21, 2023.

Firzanah, Hannah. 2021. "How Genting Highlands became Successful." Available at https://www.alvaproductions.com.my/post/how-genting-highlands-became-successful. Accessed January 22, 2023.

Genting Berhad. 2022. "Annual Report 2022: Genting Berhad." Available at https://www.genting.com/wp-content/uploads/2023/04/GENTING_AR22_FINAL.pdf. Accessed August 15, 2023.

Genting Malaysia. 2022. "Annual Report 2022: Genting Malaysia Berhad." Available at https://www.gentingmalaysia.com/wp-content/uploads/2023/04/GENM_Annual-Report-2022.pdf. Accessed August 15, 2023.

GGRAsia. 2016. "Kangwon Land to Tighten Casino Entry Rules: Report." (October 6, 2016). Available at https://www.ggrasia.com/kangwon-land-to-tighten-casino-entry-rules-report/. Accessed October 8, 2023.

Kettani, Houssain. 2010. "Muslim Population in Asia: 1950–2020." *International Journal of Environmental Science and Development* 1 (2): 143–153.

Kim, Da-ye. 2012. "Kangwon Land's Dilemma." *The Korea Times* (August 26, 2012). Available at https://www.koreatimes.co.kr/www/biz/2023/12/602_118240.html. Accessed February 11, 2023.

Kim, Eun-jung. 2016. "S. Korea Gives Casino License to Mohegan-KCC Consortium." (February 26, 2016). Available at https://en.yna.co.kr/view/AEN20160226007551320. Accessed February 20, 2022.

Kim, Hyong Chun. 1971. "Korea's Export Success, 1960–69." *Finance and Development* 8 (1): 14–21.

Kim, Hyungsok and Minwan Kang. 2015. "Developmental Direction of South Korean Casino Market." *Research in Business and Management* 2 (2): 1–10. https://doi.org/10.5296/rbm.v2i2.7776. Available at https://www.malaysiakini.com/news/15634. Accessed July 6, 2014.

Las Vegas Sun. 2001. "Booming Locals' Casino in Korea Selling Stock." (October 25, 2001). Available at https://lasvegassun.com/news/2001/oct/25/booming-locals-casino-in-korea-selling-stock/. Accessed October 7, 2023.

Lee, Choong-Ki and Kyung-Sang Kwon. 1997. "The Economic Impact of the Casino Industry in South Korea." *Journal of Travel Research* 36 (1): 52–58.

Lee, Choong-Ki and Seung-Kon Lee. 2009. "Estimating the Economic Impact of the Korean Casino Industry Using an Input-Output Model." *Korean Journal of Hospitality Administration* 18 (5): 63–73.

Lee, Donghwa, Yen-Soon Kim, Eunha Myung and Hyun Kyung Chatfield. 2018. "Integrated Casino Resort Development in South Korea: Perspectives from the Government Representatives and Industry Professionals." *Journal of Tourism & Hospitality* 7: 375. https://doi.org/10.4172/2167–0269.1000375.

Lee, Kangwook. 2010a. "Estimating the Economic Impacts of Kangwon Land Casino Resort on Gangwon Province and Other Regions in Korea." *Journal of Tourism Sciences* 34 (4): 109–126.

Lee, Young Gi. 2017. "The 'Singapore Model' in Gaming: Applications in South Korea." *UNLV Theses, Dissertations, Professional Papers and Capstones*, 3203. Available at http://dx.doi.org/10.34917/13444550. Accessed July 19, 2020.

Lee, Yunekyong. 2010b. "Estimating the Future Growth Potential of Kangwon Land Casino." UNLV Theses, *Dissertations, Professional Papers, and Capstones*, 764. https://dx.doi.org/10.34917/2040734

Lim, Seong Bae and Soon-Goo Hong. 2016. "Co-creation and Reform of State Owned Enterprise A Case Study about President Hahm's Co-creation Driven Leadership at Kangwon Land Casino and Resort." *International Information Institute (Tokyo).* Information 19 (11A): 5057–5064.

Malaysiakini. 2003. "Premier Vows No Second Casino in Msia." Available at https://www.malaysiakini.com/news/15634. Accessed June 14, 2022.

Malenovic, Marija. 2023. "Kangwon Land Inc. Receives Fine of US$2.39 Million for 182 AML Violations." (October 26, 2023). Available at https://news.worldcasinodirectory.com/kangwon-land-inc-receives-a-fine-of-us2-39-million-for-182-aml-violations-110879. Accessed October 7, 2023.

Ministry of Tourism, Arts and Culture. 2021. "History." Available at https://www.motac.gov.my/en/profile/history. Accessed October 21, 2023.

Muhibudin, Masitah, Badaruddin Mohamed and Gelareh Abooali. 2010. "A Review on Urban Tourism Development in Malaysia." In *Proceeding of Regional Conference on Tourism Research: The State of the Art and Its Sustainability*. Edited by Badaruddin Mohamed, 205–211. Available at http://eprints.usm.my/26626/1/PROCEEDINGS_RCTR_2010.pdf#page=216. Accessed October 7, 2023.

New Straits Times. 2016. "Genting Highlands Expansion Won't Involve Casino Space." (October 24, 2016). Available at https://www.nst.com.my/news/2016/10/182846/genting-highlands-expansion-wont-involve-casino-space. Accessed October 22, 2023.

Oppermann, Martin. 1992. "International Tourist Flows in Malaysia." *Annuals of Tourism Research* 19 (3): 482–500.

Phil. 2014. "Malaysia – Johor Minister Denies Reports of Hosting Malaysia's Second Casino." *Newswire* (January 13, 2014). Available at https://g3newswire.com/malaysia-johor-minister-denies-reports-hosting-malaysias-second-casino/#. Accessed July 6, 2014.

Rathakrishnan, Balan and Sanju George. 2021. "Gambling in Malaysia: An Overview." *BJPsych International* 18 (2): 32–34.
Shin, Yongseok. 2010. "Progress of and Challenges in Korean Tourism Studies." *Japanese Journal of Human Geography* 62 (6): 526–536.
TravelisMalaysia.com. n.d. "Genting Highlands." Available at https://www.travelismalaysia.com/genting-highlands/. Accessed February 20, 2022.
Wilhelm, Kelesy. 2023. "Face-to-Face: Chen Si – President, INSPIRE Entertainment Resort." *Asia Gaming Brief* (October 13, 2023). Available at https://agbrief.com/intelligence/face-to-face/13/10/2023/face-to-face-chen-si-president-inspire-entertainment-resort/. Accessed November 18, 2023.
Wu, Chung-Tong and Hae-Un Rii. 2017. "Gambling for More: Casinos as a Development Strategy for a Shrinking Korean Region." *TPR:* Town Planning Review 88 (1): 59–78.
Zoll, Patrick. n.d. "A Lesson in Gambling from South Korea." The Foreign Correspondents' Club of Japan. Available at https://www.fccj.or.jp/number-1-shimbun-article/lesson-gambling-south-korea. Accessed October 7, 2023.

Additional readings

ABS-CBN News. 2023. "Why Most S Korean Casinos Are Open to Everyone Except Koreans." (January 28, 2013). Available at https://news.abs-cbn.com/business/01/28/13/why-south-koreas-casinos-are-open-everyone-except-koreans. Accessed July 19, 2020.
France-Presse, Agence. 2007. "Lim Goh Tong, 90, Casino Founder in Malaysia, Dies." *The New York Times* (October 27, 2007). Available at https://www.nytimes.com/2007/10/27/business/27lim1.html. Accessed November 10, 2007.
Lee, Choong-Ki. 2011. "History of the Casino Industry in South Korea." *Journal of Tourism Sciences* (in Korean) 35 (1): 451–464.
Sri, Sarguna Raj, Christopher Xin Ying Tan and Ong Lee. 2020. "Malaysia: Tax, Licence and Duty Hikes Add to Operator Pressure." *G3* Magazine (October 2020). Available at https://www.theiaga.org/malaysia-tax-licence-and-duty-hikes-add-to-operator-pressure. Accessed February 6, 2021.
Suh, Hyunduk and Sung-Bum Kim. "The Macroeconomic Determinants of International Casino Travel: Evidence from South Korea's Top Four Inbound Markets." *Sustainability* 10.2 (2018): 554. https://doi.org/10.3390/su10020554.

7 Active and emerging casino tourism destinations in Southeast Asia

Malaysia was the earliest country in Southeast Asia (SEA) to formally approve a casino monopoly in 1969 in the hilltop area of Mount Ulu Kali to boost tourism development in a specific region (Chapter 6), while Singapore is a rising star which launched casino tourism in 2005 through the establishment of two iconic integrated resorts (IRs) in Marina Bay and Sentosa (Chapter 5). However, other SEA countries have also put forth effort since the 1990s to reap the potential economic benefits of casino gaming to boost the development of their own tourism industries. These include the Philippines, Vietnam, Cambodia and Laos, as well as Myanmar, which has a somewhat ambiguous status. Compared to Malaysia and Singapore, however, the business practices of casino gaming in these five countries are less transparent and commonly considered and suspected of being associated with more socially controversial/illegal market activities, like crime, drugs, questionable flows of funds and corruption.

In this chapter, the forces that contribute to developing the frontiers of casino tourism and their changes over time in the Philippines are explicated in the first section. Specifically, the reasons that led to the establishment of a state-owned gaming monopoly in 1977 and the role of government policies in the progress of the industry in the 1980s and 1990s are revealed. Besides, radical changes in the industry structure and business organization in the 2010s and onward as the result of changing market environments and government policies, and the prospects of this industry are examined. In the second section, the reasons that motivated the Vietnamese government to approve casino gaming in 1992 as part of its tourism development endeavors and the progress of this industry in Vietnam are elucidated. Also, the forces that have driven the Vietnamese government to introduce new policies that include a pilot program in 2019 with an existing casino to admit local citizens and the approval of a particular form of casino in five-star hotels that provide only electronic game venues to foreign visitors, and their implications on the prospects of casino tourism development in the country are explored.

In the third section, the particular political and geographical settings that have contributed to the development of casino tourism in Cambodia are described. Then, the forces that have led to the dramatic increase in the number of casinos in the 2010s, adverse social and environmental impacts that are being generated in the

DOI: 10.4324/9781003279884-7

related regions and actions taken by the government accordingly to prevent the escalation of the problems, and implications on the development of the industry are examined. In the fourth section, some essential features associated with the development of three casino resorts in Laos in special economic zones (SEZs) are highlighted. The business performances of these three resorts are ensured because they are endorsed by the economic reform policies of their government. However, it is noted that drawbacks in public planning and the regulatory system have negated the real benefits produced by this industry to the communities, which include increases in adverse social impacts.

In the fifth section, the role/influences of the political system in Myanmar, which is dominated by the Myanmar military on the informal operation of casinos (which are illegal by law) in some of the regions for decades, are highlighted. Besides, even though the Myanmar government legalized casino gaming in 2019, the real progress of casino tourism in this country continues to be inhibited by the long-establishing informal business routines. Finally, owing to the corrupted political and underdeveloped regulatory systems in these countries, it is concluded that the drawbacks of their legal systems and weak enforcement of their laws have indeed caused uncertainties about whether the growth of casino tourism can be ensured and sustained although reported monetary returns have been enviable in certain periods of time. In addition, the frontiers for the development of their industries will also be subjected to changes due to the policies of other governments and increasing competition in the ESEA.

Takeaways from this chapter:

1. The Philippines government initiated the development of casino tourism through the establishment of a state-owned monopoly in 1977 which is also the regulatory authority of casino gaming in the country.
2. The industrial organization of casino tourism in the Philippines experienced radical changes in the 2010s after the government discontinued the monopoly structure of the industry in 2008 and approved the entry of private operators into the market to develop IRs.
3. Government economic reform policies in 1992 launched the development of casino tourism in Vietnam, but the progress of this industry has been limited due to the political environment as Vietnam is a communist country, and there is increasing competition in the SEA region.
4. Despite various supportive policies that have been introduced by the Vietnamese government since 2017, the underdeveloped regulatory system of the country and social concerns about the adverse impacts of casino gaming both limit the prospects.
5. The Cambodian government approved casino gaming in 1995 to boost its tourism development, but the expansion of this industry in actuality has largely benefited from the banning of casino gaming in Thailand and China. Furthermore, the local citizens in the neighboring countries of Cambodia are prohibited or discouraged from entering the casinos in their own country, so this has also benefited casino tourism in Cambodia.

6. The number of casinos in Cambodia dramatically increased in the 2010s but led to serious social and environmental problems, which required radical actions by the government if the growth of casino tourism is to be sustained.
7. The progress of casino tourism in Laos was slow before 2007, but it rapidly accelerated when an existing and two new casino resorts became part of their SEZs which were established along the border with Thailand.
8. Although gambling is illegal in Myanmar, casinos have been operating openly for decades.
9. Following the revision of a Myanmar gambling law in 2019 that gave approval to foreign visitors to participate in gaming, large-scale casino tourism projects were initiated immediately afterward, despite the fact that the detailed regulations and formal casino licenses were not issued by the government.
10. Differences in the fabric of the social and political systems between countries explain for the differences in the path of development of their casino tourism and their prospects in casino tourism, as well as the attributes of the adverse social impacts being generated.

7.1 The Philippines: from a government monopoly to private operations

The Philippines is the largest archipelago country in the world and consists of over 7,600 islands.[1] The country was largely ruled by Spain between the sixteenth and nineteenth centuries and the United States of America between 1898 and 1941, and then it was occupied by Japan between 1942 and 1945. After obtaining its independence from the United States in 1946, the lack of necessary resources like fixed capital and raw materials for industrial production actually curbed its ability to make real progress, although the Philippines government promoted the modernization of the manufacturing industry in the 1950s. To boost and maintain positive growth in the manufacturing industry, low tax policies for local producers and their imports of raw materials and capital goods in the 1950s were put in place which actually resulted in a large trade deficit and had adverse impacts on the fiscal position of the government (see, e.g., Noland (1990, 80 – 81)).

Unlike the development of its manufacturing industry, its geographical location and natural resources/sceneries actually give the Philippines a competitive edge in its development as a world-renowned tourism destination. In the process, casino gaming under a government monopoly structure was first introduced at the end of the 1970s to raise tourism receipts. Indeed, casino tourism raised the foreign exchange earnings of the country. To improve the efficiency of this industry, private operators were then permitted to join the market in 2008. Following this change, the industrial organization of casino tourism in the Philippines exhibited radical changes in the 2010s, which ultimately led to the agenda to possibly privatize state-owned casinos which is still being deliberated at the time of writing of this book. Indeed, the changes in the 2010s have established the basis for the Philippines to further its development as an active casino tourism destination in the SEA.

7.1.1 Establishment of a state-owned gaming monopoly

To promote and facilitate tourism development, the Philippines government established the Philippines Tourist and Travel Association in 1952 and the Board of Travel and Tourist Industry in 1956. However, due to the underdeveloped political and legal systems, improper business activities emerged with the intention to financially profit from tourism, which included an increasing number of unlawful casinos (or gambling houses) in various tourism regions. Thus, as a "response to calls for the Philippine Government to put a stop to the growing proliferation of illegal casino operations in various parts of the country" (ICRS 2023), a state-owned and controlled gaming corporation, the Philippine Amusement and Gaming Corporation (PAGCOR), was established in 1977 to carry out measures that ousted illegal individual casinos and regulated the gaming business in the country which mainly focused on casino gaming to boost tourism development.

Under the Philippine political system, the president is empowered by the constitution to promulgate presidential decrees that directly specify the administrative requirements for the practices of the PAGCOR. In considering the potential economic benefits from the operations of casino gaming in the tourism regions, it is evident that the Philippine government exercised its authority to create favorable conditions that would facilitate the business development of the PAGCOR. These included:

> a. The exemption of PAGCOR from all import duties and taxes;
> b. The exemption of all revenues generated by PAGCOR from all forms of income taxation;
> c. Unrestricted foreign exchange controls; . . .
>
> (Lopez 2014)

In addition, as a state-owned enterprise, the distribution of the gross gaming revenue (GGR) of PAGCOR was also stated in the related legal documents in 1977. In principle, the national government was entitled to 5% of the GGR, 50% was remitted to the National Treasury, the appointed management firm for the business operations would take 10% and the remaining 35% would be retained by the PAGCOR to cover its business costs and as its profit (ibid.).

Based on the initial legal framework, the PAGCOR established the first land-based casino in an existing hotel called the Philippines Village Hotel[2] in 1979. Due to the favorable business performance of this casino, the government subsequently also approved other casino hotels and casino entertainment venues. "In 1983, by virtue of a presidential decree that consolidated and strengthened PAGCOR, eleven casino and entertainment venues were created" (Reyes 2017, 727) in and around the major tourism cities like Manlia, Cebu, Davao, Iloilo etc. Similar to the layouts of other casino tourism destinations in the 1980s like Macao and South Korea, casinos in the Philippines were set up in small to medium hotels or entertainment venues. Nevertheless, as a state-owned gaming monopoly, a large share of the gaming revenue of PAGCOR was given to the government as fiscal income. Actually, the PAGCOR made good progress in its business performance in the 1980s and 1990s. In 1999, its reported GGR was around USD220 million.[3]

To increase the publicity of the casinos and put them in the limelight, and as a result potentially increase their business turnover, the PAGCOR consolidated and rebranded its casinos in 1997 by adding "Casino Filipino" (CF) in front of the original names (e.g., Casino Filipino – Hyatt Manila; Casino Filipino – Cebu; and so on and so forth). Also, the mission of Casino Filipino clearly specified that it "is committed to helping boost the Philippines' tourism industry by providing homegrown fun and entertainment" (homepage of Casino Filipino[4]). Throughout the process of increasing the business turnover of the casinos, problem gambling issues of the local community under the practices of this industry were not found to be a critical issue, although corruption was suspected and identified at the top management level of the PAGCOR and between some of the governors over time (see, e.g., Harada and Falguera (2008, 92); Lopez (2014); and Ribaya (2012)).

To a certain extent, the gambling problems among the locals were not significant because the Philippines is a "predominantly Catholic country" (Jia 2021, 1432)[5] due to the influence of Spanish rule. To this, the Catholic religion has contributed to reducing the severity of their gambling problems, although gambling activities have long existed in the country. On the other hand, as the government's agent of this socially controversial industry, efforts from the PAGCOR also helped to lessen the potential gambling problems of the locals. In addition, direct and indirect economic benefits produced by the casino gaming business of the PAGCOR to the local workforce and other firms in the various tourism areas indeed silenced the voices of the opponents and contributed to the fundamental support of the community to sustain its operations and growth in the 1990s.

7.1.2 Transition and privatization

Casino tourism in the Philippines has continued to expand with the commencement of the twenty-first century, while its industrial structure and business organization also underwent crucial changes when the government no longer supported the monopoly position of PAGCOR in 2008. From this, the previous requirement that "business engaged in gambling operations must be affiliated with PAGCOR" (IMF 2018, 2–3) was no longer in effect, but "become subject to its regulatory powers, and operate under a PAGCOR license" (ibid., 3). Under the new arrangement, private and foreign firms can submit their proposals to the PAGCOR to apply for an individual casino license. Also, parallel to the trend of development of IRs from the second half of the 2000s, it is evident that the scale and scope of casino tourism in the Philippines have increased greatly, especially the combination of casino gaming and comprehensive non-gaming facilities within the new IRs.

For example, the soft opening of Resorts World Manila by Genting Malaysia in Newport City in 2009 marked a new era of casino tourism development in the Philippines (this IR was renamed Newport World Resorts in 2022). In addition, other privately owned IRs like Solaire Resort & Casino, City of Dreams Manila and Okada Manila were also established in the 2010s in Entertainment City, which was developed and is owned by PAGCOR in Manila Bay. Following these expansions, the Philippines has developed as a major casino tourism destination in the SEA. In terms of the GGR of the industry, the Philippines was ranked as the third-largest market in the ESEA in 2019, third to only Macao and Singapore (see Table 1.2 in Chapter 1). With respect to international tourism expenditures, the Philippines was the fourth-largest market in the ESEA (see Table 1.2 in Chapter 1) in 2019, just behind South Korea, Singapore and Malaysia.

Parallel to the rapid growth of casino tourism in the Philippines in the 2010s with the development of IRs, it has been observed that the appeal of PAGCOR casinos located in traditional and relatively small and monotonic hotels and entertainment venues to tourists has waned over time. Between 2008 and 2013, their total GGR only moderately increased and even declined in some of these years, although the number of casinos operated directly by the PAGCOR and under joint ventures with other operators increased from 37 to a record 47.[6] In contrast, the total GGR of privately owned/operated casinos (licensed casinos) increased and exceeded that of the PAGCOR casinos when they increased from 4 to 8 in the same period of time. By 2017, the GGR of 9 licensed casinos was more than three times that reaped by the 41 PAGCOR casinos. In view of this unfavorable business performance, the PAGCOR shut down some of its casinos over time. For example, the Airport Casino Filipino, which is located in proximity to the Ninoy Aquino International Airport in Manila, was closed in 2014. To explain their decision, the top management of PAGCOR indicated that "the outlook for gaming in the country is positive but that growth will be driven mostly by the private sector" (GGB 2014) in the future.

In view of the promising development of casino tourism due to private operators under the new IR business model, the issue of the dual role of the PAGCOR as a state-owned gaming monopoly and a regulator of the industry at the same time was also raised by lawmakers for discussion in 2022 (see, e.g., Simeon (2022)). To further improve the industrial organization and effective supervision of this promising industry in the country, privatization of the PAGCOR casinos was argued as the most realistic and feasible approach to exempt PAGCOR from its operation duties and conduct its role as a regulator in the market. Consequently, the chief executive officer of PAGCOR announced in 2023 that the privatization of its casinos would take place in 2025 (see, e.g., GGRAsia (2023)).

By referring to the aforementioned evidence and discussions, it is fair to argue that the approach of the Philippine government (i.e., the establishment of a state-owned monopoly) in 1977 to embed casino gaming into its tourism industry matched the contextual settings of society and the political interests/system. Despite the fact that drawbacks emerged, improvements also carried out by the government throughout the process were attempts to elevate business organizations and the overall efficiency of their casino tourism and ensure sustained contributions to the economy from the practices of this industry. Actually, the industry is taking a more active role in casino tourism development in the ESEA while the progress of this industry in the Philippines at the end of the 2000s (i.e., approving private operators to enter the market and development of the IRs) showed that the interplay between casino gaming and tourism is extending over time. Nevertheless, to ensure the long-term growth and real competitiveness of this industry, continuous improvements in the integrity of the government and increase in the regulatory power of the PAGCOR to supervise the business practices of this industry (including possible business activities) are evidently the two areas that have to be emphasized.

7.2 Progress of casino tourism in Vietnam

Vietnam has a long coastline that spans 3,260 kilometers in the east of the country, with China to its north, Laos to its west, and Cambodia to its southwest. Parallel to the success of the Communist Party of Vietnam (CPV) in gaining ruling power in North Vietnam in 1954 and the entire country in 1975, the prohibition of gambling was executed and rolled out to the territories on the same timeline. Nevertheless, approval by the CPV in the mid-1980s to reform its economy from an entirely centrally planned system to a combination of a planned and market economy compelled the government to consider casino gaming (see, e.g., Freeman (1996, 390)) as a formal business sector to boost the development of tourism in some of the selected destinations and hence increase tourist expenditures and government tax revenue.

7.2.1 Path of development under particular political inclinations

Along with the economic reform, a special license was approved for a joint venture between a major casino hotel investor from Macao and the Vietnamese government in 1992 (Ngoc 2012), which led to the opening of the first casino in Vietnam – the Do Son Casino – in 1995. Indeed, this casino resort is located in the Do Son Peninsula of a northeastern coastal city, Haiphong. Geographically, the Do Son Peninsula has a seashore landscape with natural scenic attractions, like beaches, as well as historical and cultural sites for its development as a tourism destination (see, e.g., minmax travel (2019)). Under the political inclinations of the CPV and their specific objective of approving casino gaming for tourism development, the Do Son Casino was only allowed to admit foreign visitors and overseas Vietnamese. Besides, since the practice of casino gaming remained controversial in the community, its operations maintained a low profile.

Nevertheless, in view of the increasing demand (especially from mainland China) for casino tourism in the 2000s and the expansion of this industry in the ESEA, the Vietnamese government approved five more casino licenses in the beginning of the 2000s. Of these new casinos, four[7] of them are found close to the famous coastal tourism destination of Ha Long Bay, which is in proximity to the Guangxi province of China, while another one is located in the northern city of Lào Cai, which is adjacent to the border of Yunnan province and close to its capital city, Kunming. In addition, the establishment of two more casinos was approved and they commenced operations in the 2010s. However, owing to the strict government-imposed requirements on the practices of the casinos in the various tourism destinations, any real progress of their businesses was indeed limited. Owing to the increasing competition in the ESEA and, most critically, the restrictive policies imposed by the Chinese government in the mid-2010s on the casino tourism spending of its citizens, most casinos in Vietnam reported losses in the second half of the 2010s (see, e.g., Thuy (2014) and Vietnamnet Global (2020)).

To boost spending on casino tourism, "the Vietnam Government approved a three-year pilot program" (Foster and Nguyen 2022) in 2017 for two casinos (one was in operation and another in construction at the time) to admit local Vietnamese to their gaming venues on an experimental basis from 2019 to 2022. To implement this pilot program, local patrons have "to be at least 21 years of age and have a clean criminal record to be granted entry in the casino" (Thu 2017). In addition, "[t]hey would also have to provide proof of a stable monthly income of at least VND10 million[8] (*US*$443)" (ibid.). Finally, when this program was put into operation, a daily admission fee of VND 1 million (*US*$42) or a monthly pass that cost VND25 million (*US*$10.6) was required of the local players (Barnes 2023). Although this decision generated new business opportunities for the market, real contributions to the progress of the industry were considered limited due to the specified requirements.

After tourism and casino gaming in Vietnam were hit hard by the COVID-19 pandemic in 2020 and 2021, the Vietnamese government extended the trial period for local Vietnamese to participate in casino gaming in selected casinos from 2022 to 2024 (see, e.g., Minh (2022)). At the same time, the government further allowed qualified five-star hotels in major tourism destinations to apply for the establishment of casino-type electronic games (under the category of "prize-winning electronic games") on their properties for foreign visitors.[9] Consequently, on top of the eight casino licenses through which physical table games are permitted to operate, a large number of licenses for electronic game venues were approved in various tourism destinations. "According to the Ministry of Finance, by the end of 2021, there were 63 prize-winning electronic game locations" (Foster and Nguyen 2022) in Vietnam.

7.2.2 ***Prospects and concerns***

In view of the aforementioned details associated with the process of development of casino tourism in Vietnam since the mid-1990s, it is evident that the progress of the industry was somewhat conservative before 2017 under the political climate and amid social concerns to limit its absolute scale. Considering the economic realities (mainly local employment and government tax income), however, revisions of the regulations that allowed local Vietnamese to spend money in existing casinos and the expansion of electronic gaming venues (i.e., casinos with only electronic gaming machines on their property) in five-star hotels for foreign visitors unarguably created new momentum for the growth and development of casino tourism in the country. For example, despite the severe impacts of the COVID-19 pandemic in 2020 and 2021, Casino Corona (which is allowed to admit local patrons) "served 240,560 customers from 2019 to 2021, in which Vietnamese accounted for 65%"

(Minh 2022). As a result, this casino derived VND141 billion (*US*$5.9 million) between 2019 and 2021 from local patrons, which accounted for around 5% of its total revenue, and "paid VND1.7 trillion in taxes to the government" (ibid.).

Subsequent to the policy changes in Vietnam and the positive effects on the market outlook of casino tourism in 2017, social concerns on the likely increase in adverse impacts from the expansion of this industry undoubtedly restricted the scale and scope of its development. For example, Lee et al. (2020, 1179) indicate that parallel to the expansion of casino tourism in Vietnam, its residents also "showed dissatisfaction with negative social impacts, such as higher levels of alcoholism and drug abuse, bankruptcy, prostitution and crime" and "negative environmental impacts . . . such as water pollution as a result" from the operations of the casino resorts. The findings of Lee et al. (2021, 131), who investigate the perspectives of local residents in Vietnam on casino tourism, reveal social opinions that "Vietnam should focus on promoting ecotourism and cultural tourism for its natural beauty and rich cultural heritage" aside from identifying similar adverse social and environmental impacts.

In a nutshell, the ruling Communist Party in Vietnam has the authority to legitimize a socially controversial business sector like casino gaming to boost tourism development. Politically motivated to gain social support, however, the real progress of casino tourism in the country has long been inhibited by the strict requirements of the government. Although the development of this industry saw some progress due to favorable policies in 2017, its prospects thereafter are still subjected to public concerns about the potential increase in adverse social impacts. Thus, the government continues to be prudent in its decision to support rapid increases in the scale and scope of this industry, especially in removing existing restrictions on local residents to participate in casino gaming, which may indeed curb the enthusiasm of international investors to invest in this destination.

7.3 Active expansion of Cambodia and issues

Enclosed by the borders of Vietnam to the southeast and east of the country, Laos to the northeast, Thailand to the northwest and west, and a short coastline in the south and southwest which faces the Gulf of Thailand (see Figure 1.1 in Chapter 1), Cambodia was historically "ruled by radically different political systems of organization: a traditional monarchy with feudal characteristics, a failing republic, a totalitarian communist regime, and a Vietnamese protectorate, before being governed by the UN and finally becoming a constitutional monarchy" (Cunha et al. 2015, 431). Specifically, Cambodia was the protectorate and then a colony of France between the mid-nineteenth and mid-twentieth centuries (and occupied by Japan between 1941 and 1945). After obtaining its independence in 1953, the country remained predominantly agricultural in the 1960s, although some efforts were made by the Cambodian government to industrialize its economy. Then, its economic development stagnated for two decades owing to civil war between 1970 and 1975 (which involved the United States and China) and internal political conflicts between 1975 and 1989 (as Cambodia became a protectorate of Vietnam between 1979 and 1989).[10]

7.3.1 Initial structure and dramatic changes

After the intervention of the United Nations, Cambodia restored basic social and political stability between 1989 and 1992 and then established a new government with international recognition and support in 1993. Shortly after the inauguration of its government, Cambodia introduced positive policies to revitalize its economy which included tourism development. Considering the geographical location of Cambodia and its rich historical and cultural resources, tourism is unarguably the best means to earn foreign exchange and attract the attention/interest of foreign investors to provide the needed capital funds for its development. As a matter of fact, tourism "has grown rapidly since the early 1990s and become a major contributor to domestic economic growth" (Chheang 2010, 85). At the same time, the Cambodian government also approved a casino license through an international tender. This license was granted to an investor from Malaysia and led to the development of the well-known NagaWorld Resort located in the capital city of Cambodia, Phnom Penh, in 1995.[11]

Under the political sway that the approval of casino gaming is to boost international tourism expenditures that would contribute to the development of tourism in Cambodia,[12] the regulatory framework designed by the Cambodian government for the business practices of the NagaWorld casino (and other casinos established thereafter) is similar to that of Vietnam. That is, casinos can only admit patrons with foreign passports and not local Cambodians. However, the traditions and structure of its political system which feature "corruption, nepotism and factionalism" (Cunha et al. 2015, 443) clearly undermined the enforcement of regulations for the effective supervision of the business practices of the casino. Thus, local Cambodians are actually able to enter and play in the casino due to a corrupted police force.

Unlike the incentive to approve casino gaming at NagaWorld, which targeted international tourists in general, other casino resorts established in the late 1990s were indeed strategically established for cross-border tourists from Vietnam and Thailand. As aforementioned, local Vietnam citizens were prohibited from entering the legalized casinos in Vietnam. Thus, a number of casino hotels/resorts were established in a southeastern city in Cambodia called Bavet, which borders Vietnam and is near its capital, Ho Chi Minh City. To a large extent, casinos in Bavet provide casino tourism attractions to Vietnamese visitors. On the other hand, since casino gaming is illegal in Thailand,[13] Poipet city in northwest Cambodia, which borders Thailand, gained the opportunity to become a well-known casino tourism destination for the Thais. In the 2010s, "Thai tourists account(ed) for 95% of Poipet's customers, . . . In a good year, insiders estimated Poipet's gaming revenue reach $400 million" (Cohen 2018).

Bavet and Poipet are inland cities, and developing the frontiers of their casino tourism has been largely shaped by their location, which borders Vietnam and

Thailand respectively. On the other hand, the largest casino tourism destination in Cambodia which expanded in the 2010s is located in a southwestern coastal city called Sihanoukville. This city has attractive tourism resources like historical and cultural sites, white-sand beaches and tropical sceneries. Indeed, the expansion of casino tourism in Sihanoukville was made possible by the increasing number of Chinese visitor arrivals and then Chinese capital investment in 2013 (Lim 2022). Consequently, the number of casinos in Sihanoukville has greatly increased from 15 in 2015 to 25 in 2016 and then to 30 in 2018 (Carruthers 2018). This number increased dramatically to 100 in 2019 (Rim 2023).

7.3.2 Issues and uncertainties

Under a weakly enforced regulatory system, the rapid expansion of the number of casinos in the 2010s encouraged by investors and the local government in their rush to reap the monetary benefits actually created space for questionable and illegal activities like money laundering and capital flight, and a boom in online gambling made possible within the gray areas of the legal system through the operation of the casinos (especially the small casinos with few non-gaming tourism facilities). In addition, the natural environment of Sihanoukville severely deteriorated, and various social problems also arose (see, e.g., Lim (2022); Farrelly (2022); Finney (2019); Murphy (2022); and Rim (2023)). In particular, it is widely reported that due to the rapid construction of new casinos and a wide range of tourism facilities like hotels and malls between 2017 and 2019, "[b]ricks and construction materials litter the streets, the crime rate is growing, and Sihanoukville's postcard beaches are now choked with plastic bags. The economic gap between the city's rich and poor is widening – fast" (Intrepid Travel 2020).

Actually, problems that emerged from the rapid growth of casino tourism in Sihanoukville also arose in other destinations in the 2010s but differed in severity. Indeed, a corrupt political and administrative system (including its law enforcement unit like the police department) at the outset was unarguably responsible for the escalation of these problems. Therefore, it is evident that the sustainability of casino tourism in Cambodia, especially the rapidly expanding casino segment dominated by Chinese investors is highly uncertain. To a large extent, enviable monetary statistics like those that concern foreign direct investment and business revenue highlighted by the beneficial groups from the rapid growth of casino tourism in the country are associated with both increasing social conflicts and problems. These include environmental catastrophes like water pollution and destruction of beaches in the coastal areas, abuse of local labor, crime, corruption etc. Indeed, the severity of these problems was made more clear following the outbreak of the COVID-19 pandemic in 2020 (Murphy 2022; Rim 2023).

In view of the chaotic situation under the dramatic expansion and controversial practices of casinos operating in the major tourism regions, direct government intervention was inevitable at the outset to address the severity of the problems. Fortunately, the Cambodian government did take action and promogulated a new law in 2021 which requires a casino to possess a minimum amount of capital of

USD100 million to continue/obtain its license. Thus, the "Cambodian Commercial Gambling Management Commission (CGMC) confirmed that by September 2023, the number of issued licenses to casinos operating in Cambodia had dropped to 87, after more than 200 had been operating pre-pandemic" (B2B Cambodia 2023). Nevertheless, it is worth highlighting that after this critical intervention, the number of licensed casinos in Cambodia is still the highest among the casino tourism destinations in the ESEA (see Table 2.2 in Chapter 2).[14]

Although the efforts of the Cambodian government to eliminate a large number of small and seedy casinos between 2021 and 2023 have helped to suppress the unruly expansion of its casino tourism in the 2010s, the root of the problems has yet to be addressed. That is, the integrity of the government and its capacity to institute and enforce an effective and efficient regulatory system are still awaiting to be established at the outset. Otherwise, the performance of this industry would easily be affected by changes in the policies of its neighboring countries, including the Chinese government, over time. Thus, competitiveness and sustained growth of casino tourism in Cambodia can hardly be ensured. However, given the existing realities in Cambodia, the corrupt structure and drawbacks associated with its political and regulatory system might not be easily addressed within a short period of time. Thus, the roadmap for casino tourism development in Cambodia will still be subjected to various internal and external uncertainties.

7.4 Lagging progress and reality of Laos

Like Vietnam, Laos is a communist country. Geographically, Laos has short land borders with China and Myanmar in the north and Cambodia in the south, and long land borders with Vietnam in the east and Thailand in the west. In other words, Laos is an inland country without any coastal tourism resources like Vietnam and Cambodia. Notwithstanding, natural resources like mountains, rivers and lakes, as well as historical and cultural resources offer some unique attractions for the development of tourism in Laos. Thus, following the decision of the Laos government to reform its economy in 1986 by introducing a market system (Worner 1989), tourism has been one of the industries included in the development plan of Laos to attract private investment and earn foreign exchange.

In an attempt to boost the development of tourism, the Laos government approved the first license in 1996 for a casino resort called the Dansavanh Nam Ngum Resort and Casino in the suburbs of Vientiane, the capital city of Laos which borders Thailand. Similar to Vietnam and Cambodia, local Laotians are prohibited to play in the casino. However, due to a poorly enforced and corrupted regulatory system, it is commonly known that local citizens are able to enter the casino and gamble. In practice, the accessibility and initial settings of this casino were actually not very appealing to international tourists, although the resort is located in Vientiane and in an area with some unique resources like a hilltop and lake scenery. As so, the progress of this industry in Laos between 1996 and 2006 remained small in scale and monotonic in business scope. Hence, its business performance and real contributions to tourism development in Laos have fallen short of initial expectations.

However, the decision of the government in 2003 to approve the establishment of SEZs in various regions of the country in order to further its economic reform and development (SEZO 2021) has produced new opportunities for the development of its casino tourism. To this, the first SEZ, the Savan-Seno Special Economic Zone (SSEZ), was established in 2003 and another one, the Golden Triangle Special Economic Zone (GTSEZ), was established in 2007. Since the GTSEZ is developed by the government and a Chinese company, and borders Thailand and Myanmar, this SEZ is hailed as a "new urban center and tourism" (SEZO 2021, 10) destination for Thais and Chinese investors and international tourists. Therefore, related policies are formulated to attract investment for the development of "tourism, cross-border trade and entertainment" (ibid.) in this zone. Consequently, a new casino resort, the Kings Romans Casino, was established in 2007, which was then developed into the most well-known and successful casino resort in the 2010s.

Following the establishment of the Kings Romans Casino in the GTSEZ, another casino resort, the Savan Resorts,[15] was approved in the SSEZ in 2008. Besides, as Vientiane also established its SEZ, the existing Dansavanh Nam Ngum Resort and Casino gained a new opportunity to improve its business performance. Supported by related policies and external demand, especially the flow of Thai tourists to these SEZs which border Thailand, these three casino resorts evolved into the landmarks of casino tourism in Laos. In addition, the increasing number of mainland Chinese tourists and the amount that they spend on casino tourism starting in the mid-2000s also ensure their business performance and continuous expansion. Nevertheless, taking into consideration the low transparency of the regulatory system and inefficient/weak enforcement of the laws in Laos, further development of its casino tourism continues to be inhibited by the absence of world-class international investors who have to comply with the legal requirements in their own countries to operate their overseas businesses.

In a less developed country like Laos, government supports for the development of casino tourism will unarguably increase its macroeconomic income and contribution to the development of the economy in certain areas like infrastructure construction and urbanization. However, lack of complete and effective planning and policies that support the interplay between this industry and other local tourism sectors and transfer the positive impacts to the local community may actually quash any real benefits to society. A detailed study by Sims (2017), for example, finds that while benefits derived from casino tourism development in the GTSEZ are very much controlled by the investors and government, "local residents are not simply 'missing out' on the benefits of economic growth, they are being harmed by it" (ibid., 698). Specifically, these include "socio-spatial inequalities, worker exploitation, and the large-scale forced displacement of low-income communities" (ibid.).

Besides, owing to the preferential policies granted to firms to safeguard their operations and economic benefits in the SEZs, crimes related to the practices of casino gaming could not be easily investigated by the police directly. For example, it is reported that the GTSEZ has evolved into "a haven for criminal activities including prostitution, scamming and drug trafficking, but Lao authorities are essentially powerless to stop it" (Radio Free Asia 2022). Hence, it is clear that similar to the cases in Vietnam and Cambodia, the expansion of casino tourism in Laos has largely benefited from its underdeveloped regulatory system and prohibition of casino gaming in its neighboring countries, especially Thailand and mainland China. As so, the development of the frontiers of this industry in Laos has been precarious and easily influenced by potential changes in internal and external political interests.

7.5 Ambiguous status and progress in Myanmar

Boarding China in the north and northeast and Laos and Thailand in the east and southeast is Myanmar, which gained independence from Britain in 1948. Nevertheless, political struggles and instability in the 1950s resulted in a military coup d'état (also called the Tatmadaw) in the beginning of the 1960s (Stokke et al. 2018). Actually, the military has evolved into "the foremost economic and political force" (ibid., xi) in Myanmar, and it largely controls the economic structure and business practices in the country. However, since the "military is an actor obsessed with order and stability created through the use of law, but not the rule of law" (ibid., 20), their rule has indeed undermined the economic efficiency and development of Myanmar over time. Even worse, military leaders in some regions of the country abuse their power for their own economic benefits. This is well reflected in the ambiguous status and progress of casino gaming and tourism in Myanmar.

According to Gambling Law 1986 in Myanmar, gambling is strictly prohibited in the country (DFDL 2019). However, it is well-known that some casinos and casino hotels/resorts have openly operated along their borders with Thailand and China since the 1990s and even promoted by some travel websites as "legal casinos" like the Andaman Club Resort & Casino and Treasure Island Resort & Casino[16] located in the bordering areas in the south with Thailand. In practice, the operations of these casino properties are authorized informally by the military forces and protected by their rule in different regions. Besides, the Tatmadaw is also a partner of the casino operators. On the other hand, due to the reported growth of casino tourism in the ESEA in the 2000s, individual casinos in Myanmar are multiplying and "operating illegally in some border areas and self-administered zones" (Rajah & Tann Asia 2019, 1) to profit from Thai and mainland Chinese gamblers. In this process, the practices of casino gaming (including those recognized by the Tatmadaw and the illegal ones) in Myanmar indeed foster "transnational criminal networks and are hubs for money laundering, drug supply, and trafficking in persons, for the sex trade as well as indentured labor" (Mathieson 2020, 71).

The phenomena described earlier show that Gambling Law 1986 was indeed unenforceable and could not cease the operation of casinos in the country. Thus, the Myanmar parliament approved a revision of Gambling Law 1986 in 2019 to allow the government to issue casino licenses for "the establishment and opening of casinos with specified terms" (Lincoln Legal Services (Myanmar) Limited 2019, 7). As with other countries that have justified casino gaming, "legalized casino gaming in Myanmar . . . is expected to increase tourism to the country, increase revenue generation for the State, stimulate the local economy, and create jobs for local Myanmar nationals" (Rajah & Tann Asia 2019, 1). Nevertheless, other regulations and requirements for the implementation of Gambling Law 2019 have yet to be released and the government has not yet formally approved any casino license as of 2024. Even so, it is observed that some large-scale casino resort projects have been carried out by foreign investors with the support of the Tatmadaw.

Shortly after the promulgation of Gambling Law 2019, it was observed that some Chinese investors seized the opportunity immediately to partner with the Tatmadaw to develop large mega casino resort properties in the two countryside towns of Shwe Kokko and Myawaddy, which face the Thailand border. As so, large amounts of capital funds started to flow in for the construction of infrastructures and comprehensive properties (including hotels and other hospitality and entertainment facilities) under three projects which claim to be developing three new "megacities"[17] (Tower and Clapp 2020, 5) in Shwe Kokko and Myawaddy. In principle, it is clear that these projects are attempting to establish three new casino tourism destinations instead of simply building three individual casino hotels/resorts in Myanmar.

As for the upside of these new projects, it is possible to argue that the new casinos in construction do not violate the existing gambling law because they are not yet in operation. In other words, these investments reflect the optimistic view of investors that the government will complete the detailed regulations sooner or later and then approve the casino licenses. Once the licenses are issued, the related facilities will be ready right away. However, whether these three "new cites" will really be developed into landmark casino tourism destinations in Myanmar and the ESEA or follow in the footsteps of Poipet and Sihanoukville in Cambodia will definitely depend on the contents of the legal system that will be finally formulated by the Myanmar government and actually implemented by the Tatmadaw. Besides, the prospects of casino tourism in Myanmar (if casino licenses are eventually issued) will still be subjected to changes in the policies of its neighboring countries, especially Thailand and China.

7.6 Summary

According to the United Nations,[18] the Philippines, Vietnam, Cambodia, Laos and Myanmar are "low-middle-income" countries in 2024. Besides, while the Philippines and Vietnam are categorized as developing countries, the other three are considered to be least developed countries (LDCs). Thus, it is observed that the governments of these countries are promoting spending of international tourists as a common strategy to raise foreign exchange income for their economies. In the process, the approval of casino gaming for tourism development is being viewed as a feasible public policy to not only boost monetary returns for the industry as a whole but also raise capital investment for tourism development and generate job opportunities for the locals.

However, underdeveloped legal and regulatory systems in these countries in which the enforcement of laws are further held back by corrupt governments in

Table 7.1 Corruption Perceptions Index (CPI) ranking of countries in this study (2023)

Country	*The Philippines*	*Vietnam*	*Cambodia*	*Laos*	*Myanmar*
CPI	116	77	150	126	157

Source: Transparency International. Available at https://www.transparency.org/en/cpi/2023. Accessed January 20, 2024.

The Corruption Perceptions Index ranking, which is released by Transparency International, reflects the perceived levels of public sector corruption of 180 countries/regions. While a country with a ranking of 1 means that it is the least corrupted country, 180 represents the most corrupted country in the world.

various scopes and at various degrees (which are specific to the contextual settings in the different countries) clearly inhibit the real contributions from the practices of casino tourism to their economies and societies. For example, the Corruption Perceptions Index (CPI) rankings of these five countries (as shown in Table 7.1) show that the Cambodian and Myanmar governments are among the most corrupted in the world, and the Laos and Philippines governments are also quite corrupt.[19]

Actually, the CPI rankings presented in Table 7.1 reflect the controversial environments brought about by corrupted political systems (hence underdeveloped regulatory systems) in these countries for the expansion of irregular or even illegal business activities associated with the practices of their casino tourism. Therefore, adverse social impacts have also increased in various scopes and at various degrees, as evidenced in this chapter. However, the sustainability of socially controversial business activities is always subjected to changes in the political inclinations of the governments involved and social tolerance of these activities. As discussed, the expansion and development of casino tourism in these countries (especially Vietnam, Cambodia, Laos and Myanmar) in the 2000s and 2010s were realized because casino gaming is illegal in Thailand and China, and local citizens in other neighboring countries are prohibited, restricted or discouraged from entering casinos in their own country. In fact, the policies of the Chinese and Thai governments are being revised, which will quite likely affect the development of the frontiers of these casino tourism destinations in the mid-2020s and onwards.

Specifically, while the Thai government is considering ways to recover the casino tourism expenditures of its citizens outside the country (which will be discussed in Chapter 8) and the Chinese government is restricting its people from spending on casino tourism outside the country (which will be explicated in Chapter 9), developing the frontiers of casino tourism in these five countries will definitely be affected by the policy changes of the Thai and Chinese governments. As so, revision and improvement of the controversial industrial and business organizations of casino tourism in these countries are inevitable in order to sustain their real competitiveness and development in the foreseeable future.

Notes

1 Most of these islands are very small in area. The three major groups of islands are Luzon, Visayas and Mindanao, which account for around 93% of the total land area of the Philippines. The capital of the country, Manila, is located on Luzon, which is the largest island.

2 This was an airport hotel which opened its doors in 1974 with around 300 hotel rooms and located about seven kilometers south of Manila. The Philippines Village Hotel closed in 2001 due to the redevelopment of the Ninoy Aquino International Airport.
3 The reported GGR of the PAGCOR in 1999 was 12.4 billion Philippine pesos (PHP) (Mendez and Diaz 2000). 1USD was about PHP56.2385 in 1999.
4 https://www.casinofilipino.ph/index.html
5 In terms of the size of the Catholic population, the Philippines is the third-largest Catholic country in the world, following Brazil and Mexico.
6 By referring to the data presented by the IMF (2018, 3; see Table 1).
7 Three of them are open for business while the fourth one was still under construction in 2023.
8 VND stands for "Vietnamese dong," which is the legal tender of Vietnam.
9 This is under a new decree law, No. 121/2021/ND-CP, "On the Business of Prize-Winning Electronic Games for Foreigners" promogulated in December 2021. For reference, the English version of this law is available at https://thuvienphapluat.vn/van-ban/EN/Doanh-nghiep/Decree-121–2021-ND-CP-the-business-of-prize-winning-electronic-games-for-foreigners/520415/tieng-anh.aspx#.
10 Sokty and Luyna (2013) provide an informative study on the economic history of Cambodia starting in 1953.
11 Home page of NagaWorld (https://www.nagaworld.com/nagaworld/).
12 This included an initial attempt as planned by the Cambodia government to raise the needed funds for the development of a southwest coastal province, Sihanoukville (see, for example, Yamada (2017)), which has nice beaches, tropical islands and natural resources for tourism development.
13 The case of Thailand and prospects of its casino tourism development are examined in Chapter 9.
14 In Table 2.2, the number of casinos indicated is 85, which does not include two "temporary lucky gaming licenses" (Seng 2023). This difference does not affect the argument presented here.
15 Originally, this casino resort was named Savan Vegas. According to the Laos government, a large amount of unpaid taxes of its owning firm resulted in Savan Vegas being taken over by the government in 2012 and then acquired by a Macao investor in 2016. Then, the name of this casino resort was changed to Savan Resorts. This resort was sold to a Japanese investor in 2023 (Asia Gaming Brief 2023).
16 See, for example, https://bettinggods.com/casino/does-myanmar-have-casinos/
17 These three new megacities are under the names of Yatai International City in Shwe Kokko, and Huanya International City and Saixigang Industrial Zone in Myawaddy (see, for example, Tower and Clapp (2020, 6)).
18 https://www.un.org/development/desa/dpad/publication/world-economic-situation-and-prospects-2024/
19 For comparison purposes, the CPI ranking of Singapore was 5 and of Malaysia was 57 in 2023.

References

The Philippines

GGB (Global Gaming Business Magazine). 2014. "PAGCOR Problems." (March 28, 2014). Available at https://ggbmagazine.com/article/pagcor-problems/. Accessed June 24, 2021.

GGRAsia. 2023. "Pagcor Casino Sell-off 2025, Start Price US$1.1bln: Tengco." (August 15, 2023). Available at https://www.ggrasia.com/pagcor-casino-sell-off-2025-start-price-us1-1bln-tengco/. Accessed December 2, 2023.

Harada, Yuya and Norris Z. Falguera. 2008. "Philippines' Casino Operations as a Viable Source of Revenue." *Journal of Industry and Information Science* 4 (2): 79–95.

ICRS (Integrated Corporate Reporting System). 2023. "Philippine Amusement and Gaming Corporation." Available at https://icrs.gcg.gov.ph/profiles/pagcor/. Accessed December 2, 2023.

IMF (International Monetary Fund). 2018. "Estimating Casino Revenues and Transfers for the Philippine Balance of Payments Statistics." Available at https://www.imf.org/external/pubs/ft/bop/2018/pdf/18-23.pdf. Accessed June 5, 2023.

Jia, Li. 2021. "Corporate Social Responsibility: A Case Study of Philippines Gaming Industry Agency PAGCOR." *International Journal of Law Management &* Humanities 4 (4): 1431–1443.

Lopez, Matthew Benjamin P. 2014. "Changing the Play of the Game: The Necessity to Reform the Philippine Amusement and Gaming Corporation (PAGCOR)." *SCRIBD* (August 23, 2014). Available at https://www.scribd.com/document/444340820/PAGCOR-and-the-History-of-Casino-Gaming-in-the-Philippines. Accessed December 2, 2023.

Mendez, Christina and Conrado Diaz Jr. 2000. "Pagcor Revenues Up 3.4% to P12.4 Billion in 1999." *The Philippine Star* (January 17, 2000). Available at https://www.philstar.com/business/2000/01/17/101357/pagcor-revenues-34-p124-billion-1999. Accessed December 2, 2023.

Noland, Marcus. 1990. *Pacific Basin Developing Countries: Prospects for the Future*. Washington, DC: Institute for International Economics.

Reyes Jr., Vicente Chua. 2017. "Networks of Dis (Trust) and Gaming Development in the Philippines: PAGCOR and the Entertainment City." *Pacific Affairs* 90 (4): 725–742.

Ribaya, Rio Rose. "Miriam Seeks Probe on Alleged Pagcor Bribery." (November 19, 2012). Available at https://news.yahoo.com/miriam-seeks-probe-on-alleged-pagcor-bribery.html. Accessed December 3, 2023.

Simeon, Louise Maureen. 2022. "Pagcor May Privatize Gaming Operations." *The Philippine Star* (August 31, 2022). Available at https://www.philstar.com/business/2022/08/31/2206200/pagcor-may-privatize-gaming-operations. Accessed December 12, 2022.

Vietnam

Barnes, Mark. 2023. "A Brief History of Legalized Gambling in Vietnam." *Vietnam Briefing* (July 4, 2023). Available at https://www.vietnam-briefing.com/news/a-brief-history-of-legalized-gambling-in-vietnam.html/. Accessed December 4, 2023.

Foster, Tony and Nguyen Tam Dan. 2022. "Casino Investment in Vietnam." (June 27, 2022). Available at https://www.freshfields.com/en-gb/our-thinking/knowledge/briefing/2022/06/casino-investment-in-vietnam/. Accessed December 4, 2022.

Freeman, Nick J. 1996. "Vietnam: Better Managing Reform." *Southeast Asian Affairs* (1996): 385–402.

Lee, Derrick, Pakawat Kietisaksopon and Nguyen Lan Anh. 2021. "Future Outlook of Gaming Tourism in Vietnam: Local Residents' Perspectives." In *Future of Tourism in Asia*. Edited by Anukrati Sharma and Azizul Hassan, 123–139. Singapore: Springer.

Lee, Derrick, Seongseop Kim, Heesup Han and Antony King Fung Wong. 2020. "Casino Tourism Development Is Blessing or Curse? Assessment of Casino Tourism Impacts and Suggestions for Sustainable Casino Tourism Development." *Asia Pacific Journal of Tourism Research* 25 (11): 1168–1184.

Minmax Travel. 2019. "Do Son Resort." Available at https://www.minmaxtravel.com/travel-guide/vietnam/main-cities-in-vietnam/hai-phong-city/attractive-destinations/do-son-resort/. Accessed December 14, 2020.

Ngoc, Nhu. 2012. "Casino Developers to Soon Be Able to Roll the Dice In." *Vietnam Investment Review* (October 1, 2012). Available at https://vir.com.vn/casino-developers-to-soon-be-able-to-roll-the-dice-in-16356.html. Accessed December 14, 2020.

Thu, Ngoc. 2017. "Casino Gold Rush after New Gambling Rules." *Vietnam Investment Review* (February 13, 2017). Available at https://vir.com.vn/casino-gold-rush-after-new-gambling-rules-47384.html. Accessed December 12, 2020.

Thuy, Mai. 2014. "Vietnam's Oldest Casino in the Red." *Vietnam Investment Review* (April 25, 2014). Available at https://vir.com.vn/vietnams-oldest-casino-in-the-red-27193.html. Accessed December 12, 2020.

Vietnamnet Global. 2020. "Casinos Face Continuous Losses in Vietnam." (February 24, 2020). Available at https://vietnamnet.vn/en/casinos-face-continuous-losses-in-vietnam-618981.html. Accessed December 12, 2020.

Cambodia

B2B Cambodia. 2023. "Licenced Casinos in Cambodia See Massive Decline." (September 15, 2023). Available at https://www.b2b-cambodia.com/news/licenced-casinos-in-cambodia-see-massive-decline/. Accessed December 6, 2023.

Carruthers, Marissa. 2018. "With Chinese Invasion, Sihanoukville Falls Off Radar for Europeans." (April 5, 2018). Available at https://www.ttgasia.com/2018/04/05/with-chinese-invasion-sihanoukville-falls-off-radar-for-europeans/. Accessed June 28, 2023.

Chheang, Vannarith. 2010. "Tourism and Local Community Development in Siem Reap." Ritsumeikan Journal of Asia Pacific Studies 27: 85–101.

Cohen, Muhammad. 2018. "Poipet, Cambodia's Number Two Casino Destination, Gears Up to Go Global." *Forbes* (March 16, 2018). Available at https://www.forbes.com/sites/muhammadcohen/2018/03/16/poipet-cambodias-number-two-casino-destination-gears-to-go-global/?sh=814d0511ba8a. Accessed May 28, 2018.

Cunha, Miguel Pina E, Arménio Rego, Álvaro Ferreira da Silva and Stewart Clegg. 2015. "An Institutional Palimpsest? The Case of Cambodia's Political Order, 1970 and beyond." *Journal of Political Power* 8 (3): 431–455.

Farrelly, Nicholas, Alice Dawkins and Patrick Deegan. 2022. "Sihanoukville: A Hub of Environmental Crime Convergence." Available at https://globalinitiative.net/wp-content/uploads/2022/09/GI-TOC-report_Sihanoukville_For-upload.pdf. Accessed June 28, 2023.

Finney, Richard. 2019. "Chinese Casino Pollutes Beach on Cambodian Island Near Sihanoukville." Radio Free Asia (February 22, 2019). Available at https://www.rfa.org/english/news/cambodia/pollutes-02222019154841.html. Accessed June 28, 2023.

Intrepid Travel. 2020. "The Demise of Sihanoukville: Why We No Longer Travel There." (January 15, 2020). Available at https://www.intrepidtravel.com/adventures/why-we-no-longer-travel-to-sihanoukville/. Accessed June 28, 2023.

Lim, Daryl. 2022. "The Revival of Sihanoukville, Cambodia's 'Gold Rush' City." *Think China* (July 18, 2022). Available at https://www.thinkchina.sg/revival-sihanoukville-cambodias-gold-rush-city. Accessed June 28, 2023.

Murphy, Meghan. 2022. "Foreign Casinos Have Been a Disaster for Cambodia." *The Diplomat* (July 7, 2022). Available at https://thediplomat.com/2022/07/foreign-casinos-have-been-a-disaster-for-cambodia/. Accessed June 28, 2023.

Rim, Sokvy. 2023. "Cambodia Must Check Chinese-Sponsored Casinos' Operations in Sihanoukville." (August 22, 2023). Available at https://chinaglobalsouth.com/analysis/cambodia-must-check-chinese-sponsored-casinos-operations-in-sihanoukville/. Accessed December 6, 2023.

Seng, Mengheng. 2023. "Total of 88 Gambling Licenses Issued in 2022." *Kiripost* (January 3, 2023). Available at https://kiripost.com/stories/cambodia-casino-cgmcc-gambling-licenses. Accessed May 9, 2023.

Sokty, Chhair and Luyna Ung. 2013. "Economic History of Industrialization in Cambodia." WIDER Working Paper: No. 2013/134. The United Nations University World Institute for Development Economics Research (UNU-WIDER), Helsinki. Available at https://www.econstor.eu/bitstream/10419/93708/1/773878327.pdf. Accessed June 21, 2020.

Yamada, Teri Shaffer. 2017. "Phnom Pehn's NagaWorld Resort Casino." *Pacific Affairs* 90 (4): 743–765.

Laos

Asia Gaming Briefing. 2023. "Macau Legend Gets $6 Million Less than Expected for Sale of Laos Casino Operation to Japanese Businessman." (December 21, 2023). Available at https://agbrief.com/news/laos/21/12/2023/macau-legend-gets-6-million-less-than-expected-for-sale-of-laos-casino-operation-to-japanese-businessman/. Accessed March 25, 2024.

Radio Free Asia. 2022. "Lao Authorities Seem Powerless to Stop Crime in Golden Triangle Economic Zone." (November 25, 2022). Available at https://www.rfa.org/english/news/laos/golden-triangle-11252022131629.html. Accessed June 10, 2023.

Sims, Kearrin. 2017. "Gambling on the Future: Casino Enclaves, Development, and Poverty Alleviation in Laos." *Pacific Affairs* 90 (4): 675–699.

Worner, William. 1989. "Economic Reform and Structural Change in Laos." *Southeast Asian Affairs* (1989): 187–208.

Myanmar

DFDL. 2019. "Myanmar Legal Alert: New Gambling Law Permits Casinos to Operate in Myanmar." (June 10, 2019). Available at https://asialawportal.com/myanmar-legal-alert-new-gambling-law-permits-casinos-to-operate-in-myanmar/. Accessed December 14, 2022.

Lincoln Legal Services (Myanmar) Limited. 2019. Gambling Law 2018 (Pyidaungsu Hluttaw Law No. 13/2019). (May 7, 2019). Available at https://www.lincolnmyanmar.com/wp-content/uploads/2019/05/Gambling-Law-2019-.pdf. Accessed December 14, 2022.

Mathieson, David Scott. 2020. "Non-traditional Security Challenges in Myanmar and the Greater Mekong Sub-Region." *Journal of Greater Mekong Studies* 4 (December 2020): 69–75.

Rajah & Tann Asia. 2019. "Myanmar Passes Gambling Law 2019 to Update Its Gambling Regulatory Regime, but Many Questions Remain." (September 2019). Available at https://mm.rajahtannasia.com/media/4112/2019_09_myanmar-passes_gambling.pdf. Accessed December 14, 2022.

Stokke, Kristian, Roman Vakulchuk and Indra Øverland. 2018. *Myanmar: A Political Economy Analysis*. Norway: Norwegian Institute of International Affairs. Available at https://nupi.brage.unit.no/nupi-xmlui/bitstream/handle/11250/2483349/NUPI_rapport_Myanmar_Stokke_Vakulchuk_%C3%98verland.pdf?sequence=1. Accessed July 11, 2020.

Tower, Jason and Priscilla Clapp. 2020. "Myanmar's Casino Cities: The Role of China and Transnational Criminal Networks." United States Institute of Peace. Available at https://www.usip.org/sites/default/files/2020-07/20200727-sr_471-myanmars_casino_cities_the_role_of_china_and_transnational_criminal_networks-sr.pdf. Accessed December 14, 2022.

Additional readings

Ashton, Harry and Laurie Korpi. 2008. "Asia Awakens: The Growth of Casino Tourism." Gambling Compliance Ltd. Available at https://docplayer.net/153305s77-Page-1-gamblingcompliance-asia-awakens-the-growth-of-casino-tourism-gamblingcompliance-research-team.html. Accessed June 15, 2021.

Asian Development Bank. 2017. "Tourism Sector Assessment, Strategy, and Road Map for Cambodia, Lao People's Democratic Republic, Myanmar, and Vietnam." Available at https://www.adb.org/sites/default/files/institutional-document/227186/clmv-tourism-sector-assessment.pdf. Accessed June 18, 2020.

The Philippines

Fletcher, Robert. 2023. "Pagcor Begins Switch to 'Purely Regulatory' Role." *iGaming* (September 18, 2023). Available at https://igamingbusiness.com/legal-compliance/regulation/pagcor-switch-purely-regulatory-role/. Accessed December 2, 2023.

Mendez, Christina and Conrado Diaz Jr. 2000. "Pagcor Revenues Up 3.4% to P12.4 Billion in 1999." *The Philippine Star* (January 17, 2000). Available at https://www.philstar.com/business/2000/01/17/101357/pagcor-revenues-34-p124-billion-1999. Accessed December 2, 2023.

Shead, Bob. 2017. "Gaming Industry in the Philippines Part I – Casinos." *ASEAN Briefing* (September 20, 2017). Available at https://www.aseanbriefing.com/news/gaming-industry-philippines-part-casinos/. Accessed December 2, 2023.

Vietnam

Champion Jr., Walter T. 2007. "Is Da Nang the Next Atlantic City? Why Vietnam Is the Next Sure Bet in Casino Development." *Gaming Law Review* 11 (6): 714–721.

Minh, Anh. 2022. "Let Vietnamese Gamble at Phu Quoc Casino Another 2 Years: Finance Ministry." (October 13, 2022). Available at https://e.vnexpress.net/news/economy/let-vietnamese-gamble-at-phu-quoc-casino-another-2-years-finance-ministry-4522926.html. Accessed December 4, 2023.

Cambodia

Barta, Patrick. 2012. "Cambodia Bets Casinos Can Boost Tourism." *The Wall Street Journal* (April 3, 2012). Available at https://www.wsj.com/articles/SB10001424052702304750404577317094266574870. Accessed June 18, 2016.

Muong, Vandy. 2017. "Opportune Time to Tap Growing Chinese Market in Sihanoukville." *The Phnom Penh Post* (May 25, 2017). Available at https://www.phnompenhpost.com/post-property/opportune-time-tap-growing-chinese-market-sihanoukville. Accessed June 28, 2023.

Laos

Kovathanakul, Donruetai. 2011. "Casino vs Historical Tourism Destinations Along Route No. 9 Savannakhet City." *Tourismos* 6 (2): 405–414.

Lyttleton, Chris and Pál Nyíri. 2010. "Dams, Casinos and Concessions: Chinese Megaprojects in Laos and Cambodia." In *Engineering Earth: The Impacts of Megaengineering Projects*. Edited by Stanley D. Brunn, 1243–1265. Dordrecht, Netherlands: Springer.

SEZO (Special Economic Zone Promotion and Management Office). 2021. "Special Economic Zones in Lao PDR." Ministry of Planning and Investment, Lao PRD. Available at https://laosez.gov.la/pdf/SEZ_PCL_2021_en.pdf. Accessed June 10, 2023.

Sims, Kearrin. 2016. "Laos Rolls the Dice on Casino Tourism." Available at https://www.westernsydney.edu.au/ics/news/blog/laos_rolls_the_dice_on_casino_tourism. Accessed December 13, 2023.

8 Dynamics and uncertainties of casino tourism in Japan and other potential destinations

The previous chapters have shown a clear trend in the development of casino tourism in the ESEA in the twenty-first century. There is evidence that most of the existing destinations are attempting to expand the scope and increase the scale of this industry, while other up-and-coming and potential destinations are also on the way or searching for ways to reap the benefits from the development of casino tourism. These include Japan, which passed its casino gaming law in 2018 that allows the government to grant up to three licenses for the development of integrated resorts (IRs) in the country by including casino gaming. One of these licenses was approved in 2023 to an IR project in Osaka, which will be developed through a joint venture between MGM – a Las Vegas IR operator and Orix Corporation – a Japanese company. Considering the geographical location of Japan, which has a wealth of tourism resources, including casino gaming in some of the major tourism destinations like Osaka seems to be a good idea. However, owing to the social concerns of the possible exacerbation of existing gambling problems when casino gaming becomes too easily accessible by local Japanese players within the country, the frontiers of this industry in Japan are still evolving both before its first IR opens in 2030 or so and likely afterward too.

Unlike Japan, only local referendum can determine the legitimacy of casino tourism in Taiwan, even though the Taiwanese government had approved an ordinance in 2009 that allows the establishment of tourist casinos in the outlying islands of Penghu, Kinmen and Matsu. Under this arrangement, only Matsu received the needed ballots from its voters to proceed with the development. However, owing to the delay in the required legislation at the central government level in the first half of the 2010s to enforce the ordinance and changes in the government in 2016 with the succeeding political party opposing casino gaming in Taiwan, casino tourism never materialized in Matsu. The experiences of Taiwan show that whether casino tourism can be realized in a destination is indeed a social and political, rather than a market choice at the outset. The parameters for the development of this industry can only be realized with social and political recognition. Taking into consideration that the current Taiwanese government is opposed to casino gaming, the development of casino tourism would not be feasible before another presidential election in 2028.

DOI: 10.4324/9781003279884-8

In Southeast Asia, the development of casino tourism in Thailand has long been opposed by various political and social groups. Nevertheless, the large number of Thais who visit the casino tourism destinations that border Thailand (as discussed in Chapter 7) has indeed compelled the Thai government to reassess its position on the development of this industry. Together with the desire to enhance tourism in the region, explicit discussions have been carried out in 2023 with the general expectation that casino gaming will likely be legalized in the following few years. However, the existing social structure and political and legal systems in Thailand still create uncertainties in developing the frontiers and future path of this industry.

In this chapter, the social and political reasons which led to almost two decades of debate in Japan before its Integrated Resorts Implementation Bill (hereinafter "the Bill") was passed in 2018 are explicated in the first section. Based on the Bill, the interplays between casino gaming and the local tourism industry and economy will still require time to be realized despite that one of the three potential casino licenses has already been approved to the first IR in Osaka in 2023 because the IR is not expected to open its doors until 2030. To this, the prospects of this industry in Osaka and Japan as a whole will be examined in the second section. Then, the legalization process of casino gaming in Taiwan and the issues that inhibit the development of its casino tourism will be elucidated in the third section. Finally, the forces that have led the Thai government to initiate action to legalize casino gaming in 2023 and the key issues which have to be addressed to realize the development of casino tourism in this country are explored in the final section of this chapter.

Takeaways from this chapter:

1. In Japan, advocates for the legalization of casino gaming as part of the tourism industry have mainly emphasized its economic benefits as is the case in the other discussed jurisdictions.
2. Opponents of the proposal in Japan have voiced concerns that casino gaming will magnify existing gambling problems and related adverse social impacts which are specific to Japanese society.
3. The long and slow process to approve the development of casino tourism in Japan reflects the concerns of the Japanese government on particular gambling issues and the voice of the opponents in the communities.
4. Although the market participants in Japan were excited by the approval of the first IR license in 2023, the progress and prospects of its casino tourism are still subjected to a number of uncertainties.
5. An IR operator in Japan will have to pay particular attention to the possible adverse social impacts from the practices of their casino gaming business on Japanese society if they wish to sustain their business in this country.
6. The Taiwanese government has approved an ordinance that allows the establishment of tourist casinos in the outlying islands of Penghu, Kinmen and Matsu; however, only local referendum determines the legitimacy of developing casino tourism on any one of these islands.

7. Empirical evidence in Taiwan shows that the mass community may not necessarily perceive casino tourism to have the same benefits as those promoted by the beneficial groups such as the IR operators and government, hence the proposal has been vetoed under different considerations.
8. In Thailand, the two major arguments of advocates to legalize casino gaming are to redirect the considerable volume of casino tourism expenditure of the Thais back to their own country and boost the development of its tourism.

8.1 A long process in Japan

In Japan, the first proposal to establish a modern casino was raised by a Tokyo governor in 1999 who suggested such plans for Odaiba, a reclaimed island in Tokyo Bay (see, e.g., Kamimura (2020) and Kobayashi (2001)). Similar to the usual arguments that champion the legalization of casino gaming in other jurisdictions, the rationale for this casino includes possible economic benefits for tourism and economic development, especially job opportunities for the locals when the Japanese economy was stagnant and the unemployment rate soared in the 1990s. Indeed, this idea resonated with governors in other cities and islands like Akita and Hokkaido. However, when this idea was brought to the national level for legislation, it was subsequently abandoned in 2003 without further justification. The lack of support is likely due to public concerns of the particular adverse gambling issues on Japanese society.

A decade later, a few Tokyo governors reinitiated the discussions by repackaging the proposal as a comprehensive IR (see, e.g., Kamimura (2020)) instead of simply a casino or a casino hotel. These efforts ultimately led to the approval of the National Diet of Japan (the national legislature) in 2018 to launch the development of IRs by including casino gaming. While this decision was exciting news to the market participants, the central government was slow to proceed and the unexpected outbreak of the COVID-19 pandemic in 2020 actually backtracked any initial progress. To this, the first IR is not expected to open for business until 2030. That is, it will be three decades after the idea was first raised in 1999 to legalize casino gaming as a possible impetus/attraction to promote tourism and economic development.

8.1.1 Contextual settings and reluctance to legalize casino gaming

By law, gambling businesses organized by any private organization are banned in Japan. However, it is an indisputable demand-side fact that "Japanese love gambling for leisure, to test their skill, and for a chance to win a price" (izakayaju n.d.). Thus, many gambling games have been actually made available to the Japanese in the country for decades. In practice, these gambling games are legalized as entertainment or sports, like lotteries, horse racing and motorcycle racing, as well as the most well-known Japanese pachinko and pachislot games (see, e.g., Majewski (2023)) which are part of the leisure industry. In particular, the pachinko games have evolved as part of Japanese life in the second half of the twentieth century, and they "dominate their lives, determining even the pace of their daily activities" (Tanioka 2000, 25). For a long period of time, the government has offered few policies and preventive measures to cope with the related problems of playing these games even though many Japanese have been identified as pachinko addicts (i.e., problem and pathological gamblers) and "end up causing trouble and misery to the people around them" (ibid.).

Given the high propensity of the Japanese to gamble, there have been considerable business turnovers in the markets. Also, since the games are being categorized as non-gambling activities by law, loose and insufficient regulatory frameworks to minimize gambling harm actually increase gambling activities and lead to extensive gambling problems for many Japanese at the same time. According to a comprehensive study by Takiguchi and Rosenthal (2011), the intricate relationship between the legal system and gambling issues in Japan is summarized as follows:

> [D]ue to loopholes in the law and inconsistent policies, . . . [m]ultiple government agencies are involved but none have legal responsibility for addressing the consequence of excessive gambling. There appears to be an increase in problem gambling among Japanese men who are alienated and frustrated by a deterioration in their work environment.

Actually, the annual spending of the Japanese on pachinko increased by more than double between the mid-1980s and mid-1990s from around *USD*100 billion to over *USD*200 billion (Tanioka 2000, 9), which peaked in 2005 at over *USD*300 billion (Arba 2022). Thus, the rise of adverse social impacts associated with gambling in Japan became evident in the 1990s and the beginning of the 2000s. To non-gamblers, gambling problems are unarguably a societal nightmare. However, it is important to note that despite the explicitness of gambling problems in Japanese society, it is a particular fact in Japan and the specific culture of Japan that the problems are kept low-key by politicians and communities (including the family members of the problem gamblers) because "Japan is generally understood to have a very private culture in which struggles and problems are not a matter of open discussion" (Roberts and Johnson 2016, 192).

Given these crucial facts and the particular contextual settings of Japanese society, it is not surprising that any proposal to establish a physical casino in the country (especially in the 1990s and beginning of the 2000s) would raise anxiety and discontent and the likelihood that the central government would even consider the proposal was quite low. To this, when the idea of establishing a casino in Tokyo was first raised in 1999, there were social concerns about whether the approval of a casino would further exacerbate gambling problems and related adverse impacts like family issues and crime. Therefore, abandoning the idea without much further discussion and follow-up in subsequent years is actually a kind of consensus between the interest groups in society.

Although Japanese spending on pachinko greatly declined by more than 50% between 2005 and 2021 (Arba 2022), gambling issues remain a critical problem. For example, since online casino is a common option, "one of the reasons for the reduction in pachinko play may be the accessibility of online gaming, which is also illegal in Japan" (Roberts and Johnson 2016, 182). Therefore, the idea of establishing a physical casino in the country is intuitively identical to making available another convenient form of gambling activity for Japanese gamblers, hence always generating societal concern.

8.1.2 Further efforts and final approval

As explicated in Chapter 5, the concept and practices of IR as instituted by the Singapore government in 2005 and the opening of its two IRs in 2010 clearly created the dynamics to further the development of casino tourism in the ESEA and other parts of the world. As a matter of fact, advocates of casino tourism development in Japan also revisited the idea in the beginning of the 2010s by exploring the feasibility of the development of IRs that includes casino gaming could be approved. In 2013, a new governor of Tokyo openly expressed his intention to the media that "[t]o boost the number of foreign visitors to Tokyo" (Inside Asian Gaming 2013, 46), he was seeking to obtain legal approval from the central government to establish "an integrated resort facility that combines restaurants, theaters and casinos" (ibid.). He also argued that IRs would serve as an economic instrument to revitalize the local economy, with promising new capital investment and hence job and related business opportunities for local residents and firms.

However, owing to the unexpected changes in the governorship in Tokyo between 2013 and 2016 (see, e.g., Majewski (2023)) and relatively neutral role of the successors in promoting the IR proposal, progress was stifled. Nevertheless, efforts of the advocates over time contributed to the dynamics that set the stage for the Japanese government to reassess the acceptability and feasibility of approving casino gaming as a business segment to boost the growth of the tourism industry in certain prefectures. As so, the National Diet of Japan completed discussions and passed the Act Concerning Promotion of Development of Integrated Resort Areas (hereinafter, the "IR Promotion Act") in December 2016 (see, e.g., Ishikawa et al. (2017)). As the first legal document at the national level which approves the overall direction of casino tourism development, "[t]he IR Promotion Act does not legalize wholesale casino gaming in Japan, but instead provides for the creation of a limited number of integrated resorts" (ibid.).

On the basis of the framework in the IR Promotion Act, the Japanese government continued to complete the required legal and regulatory frameworks to outline the actual structure and organization of this industry. Ultimately, the Integrated Resorts Implementation Bill (hereinafter the "Bill") was passed in July 2018, which granted the rights to the Japanese government to issue up to three licenses to interested cities in different prefectures to develop their IRs by including casino gaming as a business segment (see, e.g., GGB News (2018) and Yamaguchi (2018)). According to the Bill, interested local governments who wish to initiate an IR are required to first carry out their individual tenders to select the IR operator. Then, the local governments will submit the proposals with the IR plans to the central government for review and approval. If more than three proposals are received by the central government, the best three will be chosen under a set of evaluation criteria. Finally, the IR operators who receive a license have to submit their individual applications for casino gaming to the Gaming Control Board of Japan (see, e.g., Ishikawa et al. (2017)).

Based on the arrangements of the Bill, the industry practitioners anticipated "the first casino licenses to be issued in around the year 2020, with the first resorts to open for operation in circa 2025" (GGB News 2018). Therefore, major global IR operators from Las Vegas like MGM, Las Vegas Sands, Wynn etc. expressed their immediate interest in contributing to the development of casino tourism in Japan by cooperating with interested local governments. At the same time, some of the local interested parties like government officials and executives in various tourism-related sectors also took the initiative to learn about the intricacies of IRs in other jurisdictions like Macao and Singapore in 2018 and 2019, respectively, and explore the possible structure and real development of this industry in Japan.

However, societal concerns about the possible negative impacts from the practices of casinos in the IRs on existing gambling addiction issues (see, e.g., nippon.com (2018)) clearly prolonged the process of carrying out the actual works of IR development, like the invitation for international tenders, which was originally planned for 2018. For example, in view of the high rate of gambling addiction among the Japanese, some locals indicated their concerns and suggested that before the approval of an IR with a casino, it would be "vital for the central government and local government to enhance medical treatment related to gambling addiction, including measures to treat those addicted to pachinko and slot machine gambling" (ibid.). Thus, further discussions and exploration by the government to finalize a less controversial framework to promote IR development required a longer period of time than the local governments and potential investors had expected.

Actually, the slow progress of the central government in 2019 and the unexpected outbreak of the COVID-19 pandemic in 2020 greatly eroded the incentive of many local governors and market participants to bid for the licenses. As Blaschke (2023a) stated, although "more than 12 cities and prefectures" showed interest in competing for the three IR licenses when the Bill was passed in 2018, "only two – Osaka and Nagasaki – ultimately submitted proposals to Japan's central government" in 2022. Due to the lack of enthusiasm of the local governments to support the development of casino tourism in their prefectures, suspicion about the prospects of this industry in Japan also emerged. However, from the standpoint of the aforementioned particular social settings in Japan, the draft layout and practices of this industry may reduce its adverse impacts on society which are essential to ensure its sustained growth in the long run even though the exhilaration of many of the proponents in the country was reduced by the long process of the legalization of casino gaming and some of the demanding clauses as specified in the Bill.

8.2 A landmark decision and prospects in Japan

After more than two decades of deliberation on the legalization of casino gaming based on the concept of IRs, the central government of Japan finally completed its internal discussions and approved the first license to a joint venture between a world-renowned American IR operator – MGM Resorts International – and a Japanese financial service company – the Orix Corporation – in 2023 to launch casino tourism in Osaka. Based on the development plans announced for this IR, positive messages were clearly conveyed to the market participants (mainly the local workforce and tourism-related firms and investors). Nevertheless, another long process will ensue before the IR is ready for business in 2030. This includes the completion of the required infrastructures for the destination and other regulatory measures for the business practices. Therefore, the prospects of casino tourism in Japan will still be subjected to uncertainty of various forms. These mainly include the increasing competition from the ESEA market and changing policies of the Chinese government which discourage its citizens from spending on casino gaming abroad.

8.2.1 Excitement amid a long wait

Although the unexpected outbreak of the COVID-19 pandemic in 2020 further delayed the development of an IR in Japan, the first license was finally approved in April 2023 to MGM and Orix to develop a large-scale property with an initial investment of 1.08 trillion yen[1] in Osaka (McCurry 2023; O'Connor 2023). This announcement caused excitement in the market. Actually, this result also represents the efforts of the proponents in Osaka and its government who have long been conducting proactive research work and waiting for the opportunity to obtain the right to initiate the development of an IR (casino tourism) upon approval by the central government (see, e.g., Gardner (2014)). To a certain extent, these efforts are reflected in the selection of the location for IR development, which is to be constructed on an artificial island called Yumeshima in Osaka Bay.

As indicated in Figure 8.1, Yumeshima has good potential for developing an IR as a casino tourism destination and can generate positive spillover effects to the tourism industry and economy of Osaka. First, Yumeshima is located between the Osaka International Airport (which caters to domestic flights) and Kansai International Airport (which caters to international flights). Upon completion of the infrastructures to link Yumeshima with the existing public transportation network in Osaka, it is expected that the IR will be easily accessible by both domestic and international tourists. Besides, after arriving to the IR in Yumeshima, tourists are not only able to enjoy the comprehensive gaming and non-gaming facilities there, but many renowned tourism sites in Osaka are also accessible through convenient land transportation connections like trains and buses in less than an hour. These include Universal Studios Japan, the comprehensive shopping area of Shinsaibashi and other local landmark tourism sites like the Osaka Castle Park.

From an economic standpoint, the selection of this location will likely facilitate the efficient flow of both domestic and international visitor arrivals and hence provide considerable business opportunities to the IR operator through large-scale investments with a wide scope in both gaming and non-gaming facilities (i.e., there are benefits from internal economies of scale and scope). On the other hand, potential interplays between the IR (e.g., its non-gaming and meetings, incentives,

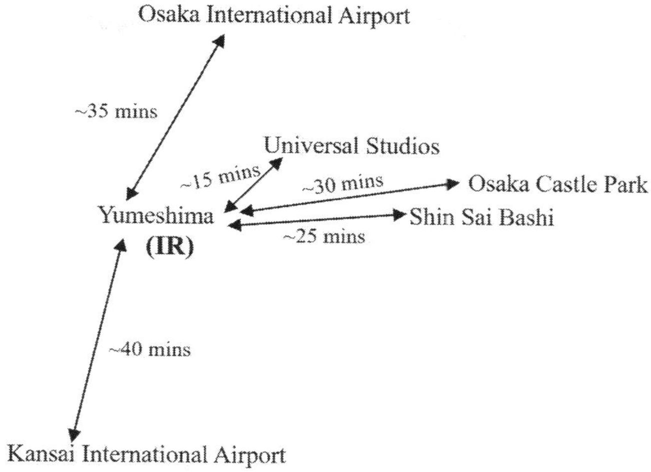

Figure 8.1 Location of IR in Yumeshima, Osaka[2]

Source: Author

conferences, and exhibitions (MICE) business segments) and the local tourism industry would also generate possible dynamics that lead to a win-win performance for most of them (i.e., benefits through the practices of external economies of scale for the local tourism industry as a whole). In addition, since the IR represents a new landmark with additional tourism attractions and facilities that include casino gaming for tourists, both new and returning visitors are expected to increase when the IR finally opens its doors.

To reap/realize the potential benefits of casino tourism through IR development in Yumeshima, however, groundwork is important and will require another few years because when the IR license was granted in April 2023, part of the land area in Yumeshima was occupied by two large-scale container terminals with a large portion of the land yet to be developed. Besides, its passenger transportation system which connects with other areas in Osaka is incomplete. Therefore, it is reasonable that the Osaka government anticipated that the IR would open in the second half of 2029 (see, e.g., Shimbun (2023)).

As aforementioned, since the approval of a casino license is subjected to a separate process upon the submission of an application from an IR operator after it obtains the IR license, this two-step process has indeed required more time by the central government than expected. Thus, the original date for the opening of the Osaka IR was postponed by about a year, which means that the earliest that this IR could be put into operation is in the second half of 2030 (see, e.g., Chan (2023) and The Japan Times (2023)). Owing to this delay, it is also estimated that the required initial investment of MGM-Orix has to be increased by almost 18% from 1.08 trillion yen to 1.27 trillion yen.[3] This is mainly "a result of the soaring price of construction materials" (Chan 2023) in the 2020s.

8.2.2 Ongoing uncertainties

As discussed in the previous sections, the long process that Japan has gone through to initiate the development of its casino tourism from when the idea was first raised in 1999 to the anticipated opening of its first IR in 2030 actually represents the consideration of the government of the potential societal opponents of the endeavor. In particular, there were public concerns that launching casinos in the country would ultimately increase the existing adverse impacts of the Japanese gambling community on society and outweigh the monetary benefits for some individual cities/prefectures. To this, the regulatory system to supervise the business practices of casino gaming and specify its interplay with other non-gaming tourism sectors are still subjected to revisions or policy tightening by the government in the following years, despite that the IR project in Osaka has already been approved in 2023.

For example, an Osaka official stated that "Japanese citizens will have to pay a 6,000 yen fee for every 24 hours they spend in the casino, with a portion of the fee earmarked for gambling addiction measures"[4] (McCurry 2023) to address the public concerns about the likely gambling problems of the local community. However, final decisions on this amount and other arrangements like the maximum number of monthly or annual visits by the Japanese to the casinos can still be under discussion before 2030. Besides, owing to the aforementioned issues associated with existing gambling problems in Japanese society (Majewski 2023), more regulations could be proposed by the government before 2030 to prevent the possible deterioration of these problems. However, business performance would undoubtedly be subjected to more uncertainties, while continuous modifications to the regulatory system may be helpful in reducing the severity of gambling problems and possible ongoing opposition to the decision of the government to launch IRs.

In addition, although Osaka and Nagasaki were the only two prefectures that submitted a proposal to the central government to bid for one of the three IR licenses, Nagasaki was finally turned down at the end of 2023. Other than concerns around the funding of the project (see, e.g., Moura (2023)), "the lack of a clear commitment to preventing gambling addiction" (Gibbs 2023) caused the Ministry of Land, Infrastructure, Transport and Tourism (MLIT) to have reservations in approving this proposal. Thus, it is clear that there is no fixed schedule for the Japanese government to grant the other two IR licenses, and the decision will be made based on whether the proposals can meet the related criteria. In view of the slow progress and delay in the development of the Osaka project, the rejection of Nagasaki and possible changes to government policies, the prospects of this industry in Japan have inevitably raised some red flags. For example, the Flexi Group (2023) commented that "Japan's foray into the realm of integrated resorts (IRs) unfolds as a saga marked by persistent ambiguity, . . . the journey thus far has been defined by a constant ebb and flow, with each step leading into an enigmatic abyss."

In principle, uncertainties associated with the lack of progress in casino gaming in Japan are mainly reflected in the market concerns about the foreseeable changes in policies of other governments and increase in competition in the region. For example, the previous chapters in this book have discussed how most governments of existing casino tourism destinations in the ESEA are continuously adjusting their policies to incentivize industry practitioners so that they devote more effort to develop their non-gaming facilities and attract foreign visitors. On the other hand, the potential increase in the number of patrons from mainland China and their expenditures are likely to slow down (although this group may not necessarily decline in number) due to the more restrictive policies imposed by the Chinese government in the beginning of the 2020s to prevent illegal capital flight from the country and gambling problems of the mainland Chinese.

In addition, since new casino tourism destinations may also be approved and developed in other parts of the ESEA in the 2020s (e.g., Taiwan and Thailand, which will be discussed in the following two sections respectively), competition for casino tourists may increase over time. To this, finding ways to ensure that the long road to realizing the first IR in Japan and the approval of two other IRs with no actual timeline can actually accommodate such a dynamic environment, and hence ensure their competitiveness in the market, definitely require proactive thoughts by the Japanese regulators and industry practitioners.

8.3 Issues in Taiwan

In Taiwan, the legalization of casino gaming has long been deemed and advocated as a tourism development strategy to promote the less-developed economies of its outlying islands like Penghu, Kinmen and Matsu. As Tsai and Gu (2005, 137) state, Taiwan's Legislative Yuan[5] "introduced for the first time in history the 'Tourist Casino Establishment Act'" (hereinafter the "Establishment Act") in 1994 in order to support and boost the development of the underdeveloped outlying islands by seizing the "potential economic and tourism benefits" (ibid.). However, in view of the opponents and challenges raised by other studies which indicate that "Taiwan did not have enough means and resources to legalize and regulate" (ibid., 138) casino gaming, as well as the desire of the small local communities in the respective islands[6] to retain their traditional and simple culture, the Establishment Act was actually set aside and not promoted to these islands.

Yet, considering the rapid expansion of casino tourism in the ESEA and its close social and economic connections with mainland China, discussions about the feasibility of legitimizing this industry in Taiwan regained public interest in the mid-2000s. The efforts of the related interest groups which included supportive politicians, industry practitioners and academics led to a new ordinance which was ultimately passed by the central government in 2009 for the establishment of tourist casinos in the outlying islands to boost their tourism development. However, the commencement of this industry has yet to be realized after more than a decade because the ordinance was vetoed twice under the local referendums in Penghu, which was the most likely island to proceed with the project with the active involvement of its governors, as well as Kinmen, which was another potential island. Besides, while the ordinance was recognized in Matsu in 2012, continuous debates and opposing voices in the local community, as well as the changing political interests in the mid-2010s have clearly put the development in limbo. Thus, the prospects of casino tourism development in Taiwan are still precarious and definitely influenced by the political interests at the central government level and social concerns at the local level (see, e.g., Wu et al. (2017)).

8.3.1 Forces leading to approval in 2009

Subsequent to the favorable progress of casino tourism in Macao after 2002 and the decision of the Singapore government in 2005 to legalize two IRs with casino gaming, advocates in Taiwan explicitly re-introduced discussions in 2007 and 2008 to legalize casino gaming[7] and made the suggestion again to the government. Compared to the motivations and legalization process in 1994, a major force that led to this round of discussions was the increasing demand of the mainland Chinese for casino tourism. Since Taiwan is a Chinese-based society with close social and cultural connections with mainland China, the increasing propensity of the mainland Chinese to spend on casino tourism not only offered attractive business opportunities to the related firms but also caught the interest of some of the politicians as a measure to raise tax revenue and revitalize the long-sluggish economy of Taiwan (including its outlying islands which are located between Taiwan and the Fujian province of China).

Given the particular demand-side factor and political interests in the 2000s, discussions ensued after the Kuomintang (KMT) Party regained power from the Democratic Progress Party (DPP) in May 2008 and a new ordinance – the Offshore Islands Development Act[8] (hereinafter the "Development Act") – was approved in January 2009. One of the most important aspects of the Development Act unarguably is the approval of casino gaming as a formal business sector to facilitate tourism development on the offshore islands aside from a comprehensive plan which outlines social, cultural and economic aspects that support the development of these islands. According to Article 10–2, the business to be legalized is literally stated as "tourist casinos" instead of simply "casinos," although it is not further specified whether those casinos are only available to tourists and not local residents. Besides, the legal entities that would operate this business are called "tourist gaming enterprises" (Article 16). Also, a more neutral Chinese term "gaming" or "playing games" rather than "gambling" is used in other articles to refer to the market activities associated with the development of this industry.

Ideologically, using the term "casino gaming" in the Development Act instead of "casino" directly in the title of this ordinance as was the case with the Establishment Act in 1994 and choosing the tag "gaming industry" indeed reflect the attempts of the legislators to convey a less negative message about the practices of this industry to the general public. Thus, it was anticipated that this would silence opposing voices in the community. Besides, considering the fact that casino gaming is indeed included, hence there is potential disagreement and opposition at the local level, Article 10–2 under the Development Act also specifies that to arrive at the ultimate decision on whether to carry out the gaming industry in any county or city, "a local referendum thereon must be held in accordance with the Referendum

Act, and more than half of the valid votes cast in the referendum must be votes of approval." However, the implementation of this crucial part of the law at the local level actually brought about unfavorable results, which held back the progress of casino tourism in Taiwan.

8.3.2 Disappointments and outlook

There was excitement in the market following the approval of the central government, and it was anticipated that Penghu Island (county) would most likely proceed to initiate the development of its casino tourism. Actually, many local governors of the Penghu Islands also showed interest and supported the proposal because the Penghu Islands are a scenic archipelago which have unique natural tourism resources like beaches. There is a common consensus that incorporating casino gaming into tourism in Penghu will produce higher value-added business from the capital investment and operations of the casino resorts, thus making Penghu more attractive to international tourists, especially those from mainland China.

Thus, a referendum was carried out in September 2009. Unexpectedly, this referendum failed to capture the interest of the masses and did not obtain "more than half of the valid votes" as specified in the Development Act for the Penghu governors to proceed with this project. In this referendum, only 18% of the 42% eligible voters who exercised their ballot voted for this proposal, while the other 24% opposed it (He 2017, 54). Nevertheless, the Penghu government initiated a second referendum in 2016. However, it was vetoed again with an even higher percentage (32%) of eligible voters who voted down the proposal and a much smaller percentage of supporting ballots (only 7%) (Hetherington 2016). Afterward, a proposal for another potential island, Kinmen, was also turned down by its referendum in 2017, even though existing tourism to the island was facing a downturn in that period of time (see, e.g., Wu and Liu (2017)). To a large extent, it can be observed that the local communities on those islands are more inclined to preserve the original features of their environment instead of pursuing revenue by introducing casino gaming to boost their tourism sector.

On the other hand, the development in Matsu was never realized, although Matsu obtained enough ballots from its referendum in 2012 to pass the casino proposal and the market initially predicted that the first tourist casino might open in 2019 (see, e.g., Chen (2013)). Other than the ongoing disputes by the opponents at the community level on whether casino gaming should ultimately be included in tourism, the required legislation for the laws and regulations at the central government level to implement the Development Act was not completed on time as expected.[9] In other words, even though the local government in Matsu obtained the required ballots to launch tourist casinos, the projects could not be launched right away before the central government completed the legislation for the Development Act. However, while the legislation process fell behind the expected timeline, the success of the DPP in regaining power from the KMT in 2016 and the opposition of its party members on gambling activities actually terminated the legislation for the Development Act.

In addition to the aforementioned experiences of Taiwan, important and realistic messages were indeed conveyed to the market even though the advocates were disappointed by these messages. First, the inclinations of the two major political parties (i.e., the KMT and DPP) crucially influenced the legalization of casino gaming at the outset, which contributed to the development of casino tourism in Taiwan. On the other hand, the mass community of a potential casino tourism destination may not necessarily perceive the same benefits from the development of this industry as those promoted by the beneficial groups like casino resort (or IR) operators and individual governors because their adopted criteria to evaluate the costs and benefits largely stem from individual and social standpoints. These are indeed different from the arguments raised by the industry practitioners and the government. In a nutshell, the reluctance of the political parties at the central government level to coordinate with (or even oppose) the industry and the lack of support at the local level of the potential destinations can hardly contribute to furthering the development of casino tourism in Taiwan in the 2020s.

8.4 Looking ahead to Thailand

Under the Gambling Act in Thailand that was promogulated in 1935,[10] any gambling activities other than horse racing and government-sponsored Thai lotteries are prohibited. Nevertheless, a number of casinos (including underground casinos) and casino resorts have been established for decades by countries that border Thailand, including Cambodia, Laos and Myanmar, to profit from the spending of Thai patrons. Therefore, legalization of casino gaming has been a topic raised by potential beneficiary groups in Thailand, including local governors and investors, at the end of the 1990s as a measure to redirect the considerable amount of funds spent by the Thais on casino tourism in the neighboring countries back to Thailand and boost the attractiveness and an image of modernity of its tourism industry and local economy. Meanwhile, some academic studies have also been conducted that examine the benefits and costs of the legalization of casino gaming in Thailand (see, e.g., Akpithanakoon (2003) and Tanawichai (2005)). In principle, there is consensus that there are economic benefits from legalized casinos to the development of potential tourism destinations in Thailand, and the considerable amount of funds spent by Thais in casino tourism outside the country could be reduced.

Nevertheless, as a Buddhist-oriented country, the Buddhist monks in Thailand are opposed to the endeavor, and there are social concerns about the possible negative and unethical impacts of casino gaming. These are the ongoing counterforces to the consideration of the government in approving this industry. Besides, long-existing problems that the general public observed, including corruption activities that involved governors, police officers and gangsters in the country, further increase the anxieties of various social groups about whether the government can formulate and implement the necessary regulatory system to supervise the business integrity of casino tourism and prevent other unsavory repercussions, including underground economic activities like drugs, money laundering, and prostitution. In addition, rivalry between the major political parties (e.g., the Move Forward and Pheu Thai parties) in the country also curbed discussions in its parliament over time.

Before 2020, the general public and international investors might have observed that the likelihood of consensus for legalization at the government level was very low, although the topic of legalizing casino gaming was occasionally brought up by some of the governors. Nevertheless, following the election in May 2023 and the formation of a new cabinet, the Thai government shared some exciting news in October 2023 to the markets that "Thailand's House of Representatives has established a new 60-member committee to consider the possibility of developing the nations' first legal casinos" (Blaschke 2023b). Considering the extent of the potential market in Thailand (including both internal and external spending on casino tourism), globally renowned IR operators like Las Vegas Sands and Genting Singapore have expressed interest in bidding for an IR license in Thailand if casino gaming is finally approved (see, e.g., Murugiah (2024) and Shriber (2023)).

Since redirecting the casino tourism expenditures of the Thai people back to their own country and tourism development have long been the two major reasons (objectives) argued for the legalization of casino gaming in Thailand, some preliminary locations considered by the Thai government include "either Pattaya and Bangkok in central Thailand; Chiang Mai and Chiang Rai in the north; Udon Thani, Ubon Ratchathani or Khon Kaen in north-east Thailand; and Krabi or Phuket in southern Thailand" (Newport 2023). While Pattaya and Bangkok (area *(a)* as shown in Figure 8.2), and Krabi and Phuket (area *(b)* as shown in Figure 8.2) are more well-known tourism destinations in Thailand which have rich natural tourism resources like tropical beaches that face the Gulf of Thailand and

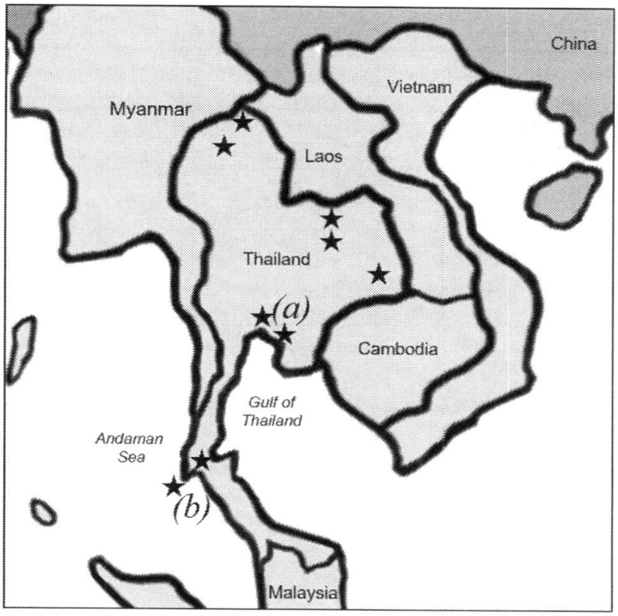

Figure 8.2 Potential IR locations (*) in Thailand

Source: Author

the Andaman Sea respectively, other possible locations under consideration are actually close to Myanmar, Laos and Cambodia. According to an initial proposal, a maximum of five cities are considered and approved to initiate the development (Blaschke 2023b).

The preliminary IR locations show that they are in line with the two aforementioned arguments to implement casino gaming as a means to boost tourism development in the respective cities, especially Bangkok and Phuket, which are well-known to Chinese tourists. Although the timeline of the legalization and organization of the potential IRs have yet to be determined (or still likely under discussion), the enthusiasm of the Thai government to drive the development of its tourism industry is unmistakable. For example, it is reported that "China and Thailand have entered into a bilateral agreement to waive visa requirements" (Interesse 2024) permanently for their citizens to travel between the two countries, starting in March 2024. As tourist arrivals from mainland China will continue to increase, there will be more motivation to drive the legalization of casino gaming. Advocates have also emphasized the economic benefits that include tax revenue, capital investment and job opportunities through the development of IRs with casino gaming.

Despite support from the Thai government, the frontier vision for the development of casino tourism in Thailand is still vague to the interest groups. For example, Cohen (2023) states in an interview with a senior industry practitioner that he believes "Thailand will do 'it' but the question is what the 'it' will look like." Actually, this view is quite pragmatic with reference to the existing social structure in Thailand and its political and legal systems. To a large extent, the business organization and practices of casino gaming in Thailand (i.e., the "it"), if ultimately approved, would not simply be outlined by the related laws and regulations but, more importantly, influenced by their enforcement. The local community and international investors are concerned about whether the integrity of the industry will be ensured due to a relatively corrupted government and regulatory system in Thailand.[11] Actually, this issue will also discourage some of the world-renowned IR operators from entering the Thai market under the regulatory requirements to retain their gaming licenses in their home countries. Last but not least, other drawbacks that affect social safety in Thailand, like "narcotics and human trafficking" (Bangkok Post 2023), definitely need to be addressed if a really competitive casino tourism destination is to be established in this country.

8.5 Summary

As discussed in the first part of this chapter, Japan has spent three decades to develop the frontiers of its casino tourism (from the first mention of a casino in 1999 to the projected opening of the first IR in 2030), which is unique to its particular social and political settings. In this long and drawn-out process, there is no question that many potential economic benefits have been forgone during the golden period for the growth of this industry in the ESEA (i.e., 2001 to 2019). Nevertheless, given the real situation of Japan like its culture, existing gambling problems that are identified within the country and its political system, the slow progress of the development of its casino tourism is not necessarily a bad thing when examined from the perspective of social stability. In addition, considering the attributes of casino gaming and its dynamic relationship with the non-gaming tourism sectors, the development of casino tourism in Japan is still a positive future endeavor. To this, proactive and efficient public policies would definitely be necessary to ensure a reasonable tradeoff between the potential economic benefits and adverse social impacts from the practices of this industry, thus ensuring its sustained growth and best contributions to the local economy and society.

As for Taiwan, its particular social connections with the Chinese community and geographical location which faces mainland China unarguably provide an attractive scenario for advocates to promote the legalization of casino gaming as a measure to boost its tourism development. However, the development cannot be realized as expected, although the establishment of tourist casinos in the outlying islands of Taiwan was approved by the central government in 2009. Unlike the process in Japan and other jurisdictions, when the rights are given to the residents in a potential jurisdiction (i.e., local level) to decide whether they would ultimately accept casino gaming and approve its legitimacy on their land through a referendum, there is no guarantee that the bill will be passed. Besides, the ongoing tug of war between the two major political parties (KMT and DPP) in Taiwan definitely hinders the prospect of the development of its casino tourism.

Finally, despite the uncertainties associated with the slow progress in Japan and loss of hope in Taiwan unless its governorship can be regained by the KMT, positive actions taken by the Thai government in 2023 to explore the possible ways to approve the development of casino tourism in the country unarguably add new momentum to the prospects of casino tourism in the ESEA. However, while the approval and legalization process will require time, the frontiers for the development of casino tourism in Thailand which include its scale and scope, as well as business routines will still be subjected to various uncertainties and possible changes even though there are some initial plans that will be released by the government in 2025 or so. Besides, creating an effective and enforceable regulatory system at the outset will be unarguably a challenging but inevitable project to the government to provide a robust foundation for the sustained growth of this socially controversial industry.

In a nutshell, by referring to the SCPE framework as specified in Chapter 1 to analyze the forces that influence the frontiers for the development of casino tourism in a destination, the three cases of Japan, Taiwan and Thailand indeed provide interesting evidence to further explicate how the same principle is being practiced differently in different jurisdictions. This is because the social, cultural, political and economic structures are quite diverse across the different countries/regions in the ESEA. Thus, whether casino tourism can be initiated in a tourism destination and the actual layout and practices of this industry vary across different countries/regions.

Notes

1 Owing to the high volatility of the foreign exchange rate of the US dollar in 2023, this amount of investment as expressed in US dollars in different newsletters varies. On average, 1USD = 140 yen in 2023.
2 At the time when this chapter was written, the infrastructure in Yumeshima and its public transportation network with other places in Osaka were still under construction. The travel time between Yumeshima and the major attractions and venues indicated in this figure are estimations of the author.
3 On average, $1USD = 140$ yen in 2023.
4 On average, $1USD = 140$ yen in 2023. This means that the fee for a 24-hour pass is around USD43. In Singapore, the daily casino entry levy is $S150, which is around USD110 (as indicated in last paragraph in Section 5.3.2 of Chapter 5).
5 The Legislative Yuan is the highest legislature in Taiwan.
6 For example, the population size of Penghu was less than 100,000 before the mid-2010s, and other major islands of Matsu and Kinmen were around 100,000 to 120,000.
7 For example, seminars/conferences were organized by some of the local academic organizations for public discussions on the possible ways to embed casino gaming into the tourism industry in Taiwan to facilitate its development and contributions to the local economy. In addition, the feasibility that the construction of casino resorts in the outlying islands would be approved was also explored.
8 The Development Act is an ordinance which requires further legislation to implement the related laws and regulations. The Chinese version of the Development Act is available at https://law.moj.gov.tw/LawClass/LawAll.aspx?pcode= A0030121 and English version is available at https://law.moj.gov.tw/ENG/LawClass/LawAll.aspx?pcode= A0030121.
9 Although a draft of the casino bill was approved by the central government in 2013 and sent to the Legislative Yuan for further review (see, for example, Inside Asian Gaming (2013)), the progress was prolonged by discussions raised by members in parliament who were not in favor of the establishment of casinos in Taiwan.
10 An unofficial English translation of this Gambling Act is available at https://docplayer.net/36825307 Gambling-act-b-e-2478–1935.html. Accessed December 18, 2023.
11 The annual rankings of the Corruption Perceptions Index, which reflect the perceived levels of public sector corruption of 180 countries/regions as released by Transparency International, can be used. In 2023, Denmark was ranked the number 1 cleanest (least corrupted) country in the world, and Thailand was ranked 108 (vs. Japan – 16, Taiwan – 28, and China –76). Available at https://www.transparency.org/en/cpi/2023.

References

Japan

Arba, Alexandru. 2022. “Net Sales Generated by Pachinko Halls in Japan from 2003 to 2021.” Available at https://www.statista.com/statistics/1171592/japan-market-size-of-pachinko-and-pachislot-market/. Accessed November 26, 2023.

Blaschke, Ben. 2023a. “Editorial: Thailand Must Learn from Japan’s IR Mistakes.” *Inside Asia Gaming* (March 30, 2023). Available at https://www.asgam.com/index.php/2023/03/30/editorial-thailand-must-learn-from-japans-ir-mistakes/. Accessed September 6, 2023.

Chan, Viviana. 2023. “Osaka IR Opening Delayed to Autumn 2030, Costs Rise $1.3bln.” *Asia Gaming Brief* (September 5, 2023). Available at https://agbrief.com/news/japan/05/09/2023/osaka-ir-opening-delayed-to-autumn-2030-costs-rise/. Accessed October 7, 2023.

Flexi Group. 2023. "Navigating Uncertainty: The Complex Landscape of Japan's Integrated Resort Odyssey." Available at https://www.flexi-news.com/post/navigating-uncertainty-the-complex-landscape-of-japan-s-integrated-resort-odyssey. Accessed December 17, 2023.

Gardner, Jessica. 2014. "Crown Executive Todd Nisbet Meets Osaka Governor in Push for Japan Casino Market." Financial Review (August 1, 2014). Available at https://www.afr.com/companies/crown-executive-todd-nisbet-meets-osaka-governor-in-push-for-japan-casino-market-20140801-j6y9p. Accessed May 2, 2023.

GGB News. 2018. "Japan IR Bill Finally Passes." (July 21, 2018). Available at https://ggb-news.com/article/japan-ir-bill-finally-passes/. Accessed July 28, 2018.

Gibbs, Erik. 2023. "Japan Rejects Integrated Resort Plan for Nagasaki." (December 27, 2023). Available at https://www.casino.org/news/japan-rejects-integrated-resort-plan-for-nagasaki/. Accessed January 1, 2024.

Inside Asian Gaming. 2013. "Governor Says Tokyo Will Have a Casino." (June 2013), 46. Available at https://www.asgam.com/mags/201306/46/. Accessed June 12, 2014.

Ishikawa, Koji, Martha Sabol and John Stapleton. 2017. "Japan Time." *Global Gaming Business Magazine* (November 25, 2017). Available at https://ggbmagazine.com/article/japan-time/. Accessed February 24, 2018.

izakayaju. n.d. "Why Is Gambling So Popular in Japan?" Available at https://www.izakayaju.ca/gambling-in-japan/. Accessed June 24, 2023.

Kamimura, Shintaro. 2020. "Japan's Tokyo Dilemma." *Inside Asia Gaming* (July 8, 2020). Available at https://www.asgam.com/index.php/2020/07/08/japans-tokyo-dilemma/. Accessed October 2, 2020.

Kobayashi, Takanori. 2001. "Ishihara in Fresh Casino Push." *Japan Today* (May 8, 2001). Available at http://www.japantoday.com/e/?content=news&id=24741. Accessed April 9, 2005.

Majewski, Antoni. 2023. "When Creativity Knows No Bounds – On the Japanese Gambling Industry." (December 8, 2023). Available at https://igamingexpress.com/when-creativity-knows-no-bounds-on-the-japanese-gambling-industry/. Accessed December 14, 2023.

McCurry, Justin. 2023. "Japan Approves Building of First Casino." (April 14, 2023). Available at https://www.theguardian.com/world/2023/apr/14/japan-approves-building-of-first-casino. Accessed May 2, 2023.

Moura, Nelson. 2023. "Financial Concerns Behind Gov't Decision to Bar Nagasaki IR Proposal." *Asia Gaming Brief* (December 27, 2023). Available at https://agbrief.com/news/japan/27/12/2023/financial-concerns-behind-govt-decision-to-bar-nagasaki-ir-proposal/. Accessed January 1, 2024.

nippon.com. 2018. "Pachinko and Slot Machine Addiction in Japan." (August 2, 2018). Available at https://www.nippon.com/en/features/h00252/. Accessed December 14, 2020.

O'Connor, Devin. 2023. "MGM Resorts Wins First Casino License in Japan with $8.1B Osaka Bay Resort." (April 14, 2023). Available at https://www.theguardian.com/world/2023/apr/14/japan-approves-building-of-first-casino. Accessed May 2, 2023.

Roberts, Jennifer and Ted Johnson. 2016. "Problem Gambling: How Japan Could Actually Become the Next Las Vegas." *UNLV Gaming Law Journal* 6 (2): 176–198.

Shimbun, Yomiuri. 2023. "Osaka Casino Resort Plans Likely to Be Approved." *The Japan News* (April 12, 2023). Available at https://japannews.yomiuri.co.jp/politics/politics-government/20230412-103124/. Accessed May 2, 2023.

Takiguchi, Naoko and Richard J. Rosenthal. 2011. "Problem Gambling in Japan." *Electronic Journal of Contemporary Japanese Studies*. Available at https://japanesestudies.org.uk/articles/2011/Takiguchi.html. Accessed July 18, 2022.

The Japan Times. 2023. "Osaka Signs Deal to Open Casino Resort in Fall 2030." (September 29, 2023). Available at https://www.japantimes.co.jp/business/2023/09/29/osaka-casino-resort-deal/. Accessed October 7, 2023.

Yamaguchi, Mari. 2018. "Japan Parliament OKs Law to Allow Up to 3 Casino Resorts." (July 20, 2018). Available at https://lasvegassun.com/news/2018/jul/20/japan-parliament-oks-law-to-allow-up-to-3-casino-r/. Accessed July 28, 2018.

Taiwan

Chen, Kobe. 2013. "Modernizing Matsu: From Fortress to Casino Resort." *Taiwan Panorama* (June 2013). Available at https://www.taiwan-panorama.com/Articles/Details?Guid=dfcb470f-17bb-4e0e-9439-21b1d32586a4&langId=3&CatId=7. Accessed June 12, 2022.

He, Miao. 2017. "A Comparative Study of the Political Economy of Asian Casino Gaming." *Ph.D. Thesis*, Faculty of Business Administration, University of Macau. Available at https://library2.um.edu.mo/etheses/b3954137x_ft.pdf.

Hetherington, William. 2016. "Penghu Says No to Casinos for Second Time." *Taipe Times* (October 16, 2016). Available at https://www.taipeitimes.com/News/front/archives/2016/10/16/2003657252. Accessed July 14, 2023.

Inside Asian Gaming. 2013. "Taiwan Releases Casino Legislation." (May 7, 2013). Available at https://www.asgam.com/index.php/2013/05/07/taiwan-releases-casino-legislation/. Accessed June 12, 2022.

Tsai, Henry and Zheng Gu. 2005. "Penghu Gambling Legalization: A Viable Alternative for Taiwan." *Gaming Law Review* 9 (2): 136–143.

Wu, Cheng-ting and Li-jen Liu. 2017. "Kinmen Nixes Casino Development." *Taipei Times* (October 21, 2017). Available at https://www.taipeitimes.com/News/front/archives/2017/10/29/2003681240. Accessed July 14, 2023.

Wu, Jeanette Mei-Ian, Henry Tsai and Jin-Soo Lee. 2017. "Unraveling Public Support for Casino Gaming: The Case of a Casino Referendum in Penghu." *Journal of Travel &* Tourism Marketing 34 (3): 398–415.

Thailand

Akpithanakoon, Yuwaluck. 2003. "The Study on the Establishment of a Legal Casino in Thailand." *Master Thesis*, Assumption University. Available at https://repository.au.edu/bitstream/6623004553/5916/1/AU-Project-Report-Fulltext-167406.PDF. Assessed December 17, 2023.

Bangkok Post. 2023. "Casino Plans Need Rethink." (November 11, 2023). Available at https://www.bangkokpost.com/opinion/opinion/2682829/casino-plans-need-rethink. Accessed November 12, 2023.

Blaschke, Ben. 2023b. "Thailand Establishes New House Committee to Study Prospect of Legal Casinos." *Inside Asia Gaming* (October 30, 2023). Available at https://www.asgam.com/index.php/2023/10/30/thailand-establishes-new-house-committee-to-study-prospect-of-legal-casinos/. Accessed November 12, 2023.

Murugiah, Surin. 2024. "Genting Singapore Bay Bid for Thailand Casino – Maybank IB." *The Edge Malaysia* (January 7, 2024). Available at https://theedgemalaysia.com/node/696471. Accessed January 20, 2024.

Newport, Alastair. 2023. "Thailand Gets Closer to Legalised Gambling." (November 2, 2023). Available at https://www.ttgasia.com/2023/11/02/thailand-gets-closer-to-legalised-gambling/. Accessed November 12. 2023.

Shriber, Todd. 2023. "Thailand House of Reps Forms Casino Committee." (October 30, 2023). Available at https://www.casino.org/news/new-committee-will-examine-issue-of-thailand-casinos/. Accessed January 20, 2024.

Tanawichai, Siriluck. 2005. "In Pursuit of Development: Casinos and Local Government in Northern Thailand." *Ph.D. Thesis*, The University of Waikato. Available at https://researchcommons.waikato.ac.nz/handle/10289/13208. Accessed December 17, 2023.

Additional readings

Japan

GRAsia. 2018. "Japan's IR Bill Passes into Law, Main Work Starts Now." (July 20, 2018). Available at http://www.ggrasia.com/japans-ir-bill-passes-into-law-main-work-starts-now/. Accessed July 22, 2018.

GGRAsia. 2022. "MGM-Orix Not Mulling Osaka IR Escape Clause: Execs." (March 17, 2022). Available at https://www.ggrasia.com/mgm-orix-not-mulling-osaka-ir-escape-clause-execs/. Accessed April 2, 2022.

O'Connor, Devin. 2018. "Japan Integrated Resort Committee Details Casino Site Selection Process." (October 9, 2018). Available at https://www.casino.org/news/japan-integrated-resort-committee-details-casino-site-selection-plan/. Accessed December 14, 2020.

O'Connor, Devin. 2021. "Japan Integrated Resort Bill BECOMES Law July 19, Bidding Begins in October." (July 15, 2021). Available at https://www.casino.org/news/japan-integrated-resort-casino-bill-becomes-law-july-19/. Accessed July 18, 2021.

Penn, Michael. 2020. "Japan IRs Have a New Schedule." *Asia Gaming Brief* (October 12, 2020). Available at https://agbrief.com/news/12/10/2020/japan-irs-have-a-new-schedule/. Accessed December 5, 2020.

Philander, Kahlil, Brett Abarbanel and Bo Bernhard. 2017. "Socio-Economic Impacts of Japanese Integrated Resorts." International Gaming Institute, University of Nevada, Las Vegas. Available at https://static1.squarespace.com/static/600a2a31d2a8133c35992c23/t/64878d6e0c646478eed327af/1686605170583/JapanSocialEconomicImpactsReport.pdf. Accessed May 10, 2021.

Tanioka, Ichiro. 2000. *Pachinko and the Japanese Society*. Osaka, Japan: Institute of Amusement Industries, Osaka University of Commerce. (English edition by: Osawa, Minae, Madeleine Vedel and Okamoto Miki).

Thompsett, Louis. 2022. "Asia Round-up: Osaka & Nagasaki Casinos Face Delay; Ronnie O'Sullivan Signs Deal." *Gambling Insider* (December 9, 2022). Available at https://www.gamblinginsider.com/news/19443/asia-round-up-osaka-nagasaki-casinos-face-delay-ronnie-osullivan-signs-deal. Accessed October 7, 2023.

Taiwan

Casino.org. 2017. "Voters on Taiwan's Kinmen Islands Overwhelmingly Reject Call for Legal Casinos." (October 30, 2017). Available at https://www.casino.org/news/voters-on-taiwans-kinmen-islands-strongly-reject-casino-proposal/. Accessed July 14, 2023.

Conneller, Philip. 2018. "Legal Taiwan Casinos Would Bridge $9.7 Billion Tourism Deficit, but Government Opposed." *Casino.org* (October 10, 2018). Available at https://www.casino.org/news/legal-casinos-in-taiwan-would-bridge-9-7-billion-tourism-deficit-but-government-opposed/. Accessed July 15, 2023.

Hioe, Brian. 2017. "Penghu Votes against Casino Development." *New Bloom* (October 17, 2016). Available at https://newbloommag.net/2016/10/17/penghu-casino-referendum/. Accessed July 14, 2023.

Inside Asian Gaming. 2013. "Taiwan Releases Casino Legislation." (May 7, 2023). Available at https://www.asgam.com/index.php/2013/05/07/taiwan-releases-casino-legislation/. Accessed July 14, 2023.

Thailand

Blaschke, Ben. 2023. "Thailand Establishes New House Committee to Study Prospect of Legal Casinos." *Inside Asia Gaming* (October 30, 2023). Available at https://www.asgam.com/index.php/2023/10/30/thailand-establishes-new-house-committee-to-study-prospect-of-legal-casinos/. Accessed November 12, 2023.

Cheng, Daniel. 2023. "The Wheel Is Turning for Casino 'Entertainment Complexes' in Thailand." *Asia Gaming Brief* (November 3, 2023). Available at https://agbrief.com/news/thailand/03/11/2023/the-wheel-is-turning-for-casino-entertainment-complexes-in-thailand/. Accessed January 2, 2024.

Cohen, Muhammad. 2023. "Thailand Casino Landscape Forming as Details Remain Fuzzy." *iGaming Business* (September 5, 2023). Available at https://igamingbusiness.

com/legal-compliance/regulation/thailand-casino-landscape-forming-as-details-remain-fuzzy/. Accessed January 2, 2024.

Highstakes. 2023. "Thailand's Casino Landscape to See New Developments." (September 13, 2023). Available at https://highstakesdb.com/news/high-stakes-reports/thailand-s-casino-landscape-to-see-new-developments. Accessed November 12, 2023.

Interesse, Giulia. 2024. "China and Thailand to Waive Visa Requirements Permanently from March." *China Briefing* (January 2, 2024). Available at https://www.china-briefing.com/news/china-and-thailand-to-waive-visa-requirements-permanently-from-march/. Accessed January 20, 2024.

9 Prospects of casino tourism in East and Southeast Asia

After the rapid growth and development of casino tourism in the ESEA in the 2000s and 2010s, there have been inquiries from the markets about the sustainability and future roadmaps of this industry. To evaluate the prospects of this industry in the foreseeable future, the role of the "China factor" and its changes must be examined. This is because Chinese outbound tourists have become the largest source of visitor arrivals to most of the casino tourism destinations in the ESEA and their spending has contributed to a large portion of the business turnovers in these destinations. Thus, changes in the policies of the Chinese government and consumption behavior of the mainland Chinese tourists when they visit the casino tourism destinations are unarguably significant forces that will alter the existing path of development and business organization of this industry.

Parallel to the impacts from the changing role of the "China factor," the development of casino tourism in the various ESEA destinations is likely to go through a process of expanding the mass tourist market instead of continuing to rely on the immense spending of heavy (or "VIP") gamblers in a loosely regulated environment that is practicing between law and custom. To compete for the mass patrons, there is an increasing trend of advocating for the expansion of non-gaming facilities and attractions in these destinations. In addition, to cultivate and sustain the competitive edge of a destination, effectively connecting the non-gaming facilities of IRs and local tourism resources in the destination is also an essential part of this trend. Overall speaking, casino tourism in the ESEA is clearly entering a stage of adjustment and transformation following a period of rapid growth and development. In this stage, changes in the political inclinations on the expansion of casino tourism will also give rise to uncertainties in its prospects, while new dynamics will likely be generated in the markets for further development of this industry.

In the first section of this chapter, the reasons for the Chinese government to amend its criminal law in 2021 which prohibits anyone from organizing local Chinese citizens to participate in casino gaming outside the borders of mainland China are identified. To this, the impacts on the prospects of casino tourism in the ESEA are evaluated. Aside from government policies, changes in the consumption behavior of the younger Chinese in visiting casino tourism destinations and the related business implications are examined. In the second section, arguments and evidence for the increasing emphasis on the development of non-gaming elements in the

DOI: 10.4324/9781003279884-9

casino tourism destinations are discussed. To this, the significance of achieving effective interplays between non-gaming facilities established in the IRs and the local tourism resources/attractions of the respective casino tourism destinations is explained.

In the third section, the impacts of changes in the environmental factors and new dynamics of IR development as indicated in the first two sections on the prospects of casino tourism in the ESEA are examined. Parallel to the positive dynamics, uncertainties generated from the possible changes in government policies (especially those in China) to the prospects of this industry are uncovered. To conclude, the forces that lead to different practices and ongoing changes of this industry in the different jurisdictions in the ESEA are recapped. Also, the importance of the role taken by a government at the outset to develop the frontiers for sustaining the growth and development of this socially controversial industry is revisited.

Takeaways from this chapter:

1. The amendment of the Chinese government of its criminal law in 2021 curbed the rapid expansion of the industry, as shown in the 2000s and 2010, and altered the path of development of this industry to emphasize instead on cultivating the mass gaming markets.
2. To boost the development of the mass market, the industry operators need to make proactive and ongoing innovative efforts not only to address the changing tourism consumption behaviors of the Chinese millennials and Generation Z (Gen Z) but also to provide different ways to encourage their spending.
3. Non-gaming elements will play a more important role in cultivating/strengthening the competitiveness of a casino tourism destination, hence sustaining its development.
4. The composition of non-gaming elements in a destination should not be restricted to the facilities established within the IRs but extended to tourism resources which are endowed and unique to the destination. Effective interplays between these elements have to be ensured.
5. Due to the innovative powers of the IR operators, the success of the IRs and their ongoing development will sustain the momentum for the vitality of casino tourism in the ESEA.
6. The prospects of casino tourism in the ESEA will be subjected to uncertainties because of the potential changes in political interests/inclinations, and hence the policies for the ongoing practices and progress of this industry.
7. The ESEA governments need to have an effective role at the outset to ensure the sustained growth and development of casino tourism in the ESEA.

9.1 Dynamics from changes in the "China factor"

Throughout the process of developing the frontiers of casino tourism in the various ESEA destinations in the 2000s and 2010s, the evidence and discussions presented in the previous chapters show that other than the internal social, cultural, political and economic (SCPE) forces which are specific to those countries, the

"China factor" has specifically evolved as an important external force that influences/changes these frontiers. Indeed, the significance of the "China factor" will continue to increase. With a population of over 1.4 billion individuals in 2023[1] and rising income per capita, the potential number of Chinese outbound tourists in the following decades will definitely provide considerable business opportunities for the development of existing and potential casino tourism destinations in the ESEA. To seize these opportunities, business (especially IR) operators should make positive and innovative efforts that enhance the attractiveness of their properties and service quality over time.

However, given that the Chinese government has outlawed gambling and casino tourism, trips taken by the mainland Chinese outside the country have been often identified and suspected of being associated with various socially and politically undesirable activities like over gambling, usury, illegal flow of funds outside the country etc. (for, e.g., evidence shown in the previous chapters for Macao and other destinations in the SEA), and so in the first two decades of the twenty-first century, the Chinese government has tightened its policies and enforcement of the laws in 2021 to cope with the problems. As a response, many governments of the casino tourism destinations in the ESEA are indeed revising their policies to increase supervision and integrity to reduce the impacts of the Chinese government policy, hence ensuring real competitiveness and sustaining the growth of their destination.

Other than the challenges produced because of the changes in the policies and actions taken by the Chinese government, it is also observed that the attributes and preferences of Chinese patrons for casino tourism are evolving in the 2020s. Specifically, the motivations to take part in casino tourism and consumption behavior of the younger Chinese (i.e., Chinese millennials and Gen Z[2] and, in the future, Generation Alpha) for casino tourism are clearly different from those of the older generation (see, e.g., Huang (2023) and Meneses (2022)). Actually, an increasing number of the younger Chinese are not primarily visiting casino tourism destinations like Macao and Singapore (as well as Las Vegas) for casino gaming. Under such dynamics, the composition and business organization and routines of casino tourism in the ESEA will inevitably need to undergo continuous and radical changes ahead.

9.1.1 Changes in Chinese government policies

The success of the Chinese government in reforming and advancing the development of its economy means that China has now surpassed the gross domestic product of Japan in 2010 and is now the second-largest economy in the world behind the United States of America. Parallel to the increase in national income, the Chinese government also introduced more flexible visa policies at the beginning of the 2000s, which allowed mainland Chinese citizens to travel outside of China.[3] Thus, the number of Chinese outbound tourists who are traveling for personal reasons has increased significantly from less than seven million person-times in 2001 to over 162 million person-times in 2019,[4] which places China as the largest source of outbound tourists in Asia.[5] To this, the increasing spending of mainland Chinese tourists is an explicit force that contributes to the expansion of the tourism industries in various countries/regions in the ESEA.

At the same time, the group of mainland Chinese who are benefiting from the successful development of their economy and spend on casino tourism are ever increasing over time. The spending of this group has clearly excited a number of the neighboring countries of China, as this has facilitated the expansion and establishment of new casino tourism facilities to reap the economic benefits. Indeed, it is estimated that in the 2010s, "[m]ost of the 340 or so casinos in South-East Asia were set up with the explicit aim of luring Chinese gamblers. And many have enjoyed a bonanza: an estimated 1trn yuan ($144bn) in gambling money leaves the Chinese economy every year" (The Economist 2022, 51). As for the increasing spending of the mainland Chinese on casino tourism in the 2010s, it is found and often suspected that a significant portion of the money is actually associated with public and private sector corruption and illegal flow of funds, which include capital flight from China to counter the capital control policy of the Chinese government.

Owing to the changes in the tolerance of the Chinese government of the irregular and illegal activities associated with casino tourism undertaken by its people outside the country and to resolve the related social and political problems found in the country, policies have been made that restrict the expansion of casino tourism activities undertaken by the mainland Chinese. In 2020, the Chinese government conducted a radical amendment of Article 303 of the Criminal Law of the People's Republic of China, which was put into effect on March 1, 2021.[6] This amendment specifies that establishing casinos in mainland China will face a maximum of five years of imprisonment and a concurrent fine, and serious cases face an imprisonment term of five to ten years and a concurrent fine. Most critically, anyone who organizes local citizens in mainland China for gambling outside the borders of mainland China will be penalized with the same penalties of establishing casinos in mainland China (i.e., imprisonment and concurrent fine).

The provisions of this article clearly show the legal requirements for anyone who promotes tourism (including foreign operators) to the local Chinese citizens in mainland China. The promotion cannot involve any casino gaming activities as of 2021, which includes the IRs in Macao. As Moura (2021a) commented,

> [A]lthough analysts have considered that the changes target mainly gaming operations in the Philippines, Cambodia, Myanmar, Laos and Vietnam the wording of the new amendment's establishes that cope of action to gambling "outside the country (borders)", leaving the possibility for Macau gambling operations to be covered by it.

Indeed, the Chinese government has not only conveyed a clear message to both society and the markets but also enforced a law immediately that cracks down on related business activities conducted in mainland China (see, e.g., AGBrief Editorial (2022); Moura (2021b) and Purbrick (2024)). Thus, some of the existing business routines, especially the widely adopted gaming-agent business model (e.g., the traditional "Macao model" as explicated in Chapter 4) in various destinations to promote outbound casino tourism trips to the mainland Chinese, are likely to cease over time.

Actually, radical changes in the business organization and routines in casino tourism in Macao after 2023 (see Chapter 4) provide possible signals of the potential changes in the frontiers of casino tourism in the ESEA. In particular, the trend of the industry in the 2000s and 2010s to reap the revenue from the generous spending of Chinese high-rollers and heavy gamblers on casino tourism trips will likely fade out. In the process, it would be reasonable to anticipate that the dynamics for the development of casino tourism will change to accommodate the mass gaming segment, with increasing emphasis on better-regulated gaming activities and a larger share of non-gaming tourism activities for Chinese patrons. Given the diverse contextual settings of the various ESEA countries, this direction is imminent, although their changes will not be at the same pace and are specific to their own reality. For example, the Cambodian government announced in 2024 that "it is planning to halt the expansion of gambling within the country" (Wilhelm 2024). In principle, this decision is "likely be a bid to cozy up to its main regional friend China and get visitation back" (ibid.).

9.1.2 Changing consumption behavior of Chinese outbound tourists

While policy changes in China would likely compel the more developed casino tourism destinations like Macao, Singapore, Malaysia and South Korea, as well as the Philippines and the up-and-coming player, Japan, to emphasize more on expanding the mass market in the future, the ongoing changes in the consumption behavior of the Chinese visitors during their casino tourism trips will be an essential force that requires industry operators to make crucial and proactive changes to existing business components (especially non-gaming facilities). Actually, it is observed that Chinese outbound tourists are younger over time. For example, Bao et al. (2019, 563) refer to a World Tourism Cities Federation Report in 2015 and point out that "the majority of outbound mainland Chinese travellers over the past 10 years are younger (with those born after 1980s and 1990s accounting for 68%)," and this trend is expected to definitely continue.

In the 2010s and 2020s, this younger group of Chinese outbound tourists was largely described as Millennials who have a higher education than their parents and many of them are working in the modern business sectors with individual expertise. On the other hand, this younger group of travelers will increase in number with the addition of Gen Z, who were born between 1997 and 2012 and are now of age for traveling. Owing to the changes in the social and technological environments (especially the rapid pace that China is evolving into a digital society/economy), when Gen Z matures and their education level continues to rise, their motivation to visit a casino tourism destination and spending patterns at such destinations will be quite different from those of their parents and grandparents. Also, although online gaming (without wagers) and gambling (with wagers) activities have been commonly recognized among the Chinese millennials and Gen Z, they may not be overly enthusiastic toward traditional and physical casino games when visiting a casino tourism destination. Thus, even if they play these casino games, they may not wager very large amounts of money.

For example, Meneses (2022) shows that the Chinese millennials exhibit quite different characters and lifestyles, including their choice of leisure and traveling activities, as compared to the previous generations. An industry financial analyst comments that in terms of gaming, the Chinese "millennials attending casinos spend more time and money on food, drink and entertainment than they do on gambling" (ibid., 62). Besides, it is further emphasized that "a millennial is more interested in skill-based games 'such as fantasy sports and poker and is more familiar with online gaming'" (ibid.). According to these comments and study results, it is clear that in the foreseeable future (i.e., the mid-2020s and onward), industry practitioners (IR operators) will need to make proactive and innovative efforts to capture the younger Chinese outbound tourist market who have potentially high purchasing power. Indeed, these efforts are essential to sustain the growth and development of this industry. Additionally, non-gaming attractions and facilities will expand over time to meet and stimulate the demand of mass Chinese outbound tourists, while the composition of the gaming devices (i.e., product mix) in the casino tourism destinations will be subjected to continuous changes.

9.2 Reshaping the role and composition of non-gaming elements

Parallel to the dynamics generated from the aforementioned changes based on the "China factor," better and more effectively regulated markets which focus on providing casino gaming as a kind of entertainment and attraction to the mass tourists instead of simply attempting to earn a large amount of revenue from the gambling of high rollers or heavy gamblers over time have been the trend. This is likely to continue in most ESEA destinations in the 2020s and onward. In the process, the continuous revamping of their industrial organization and business routines to align with this trend in the ESEA as a whole will be inevitable even though some of the less developed casino tourism destinations in the SEA may need more time and face more uncertainties in their search for the right amount of change. Indeed, these advancements are important for all of the destinations to sustain their growth and ensure positive and long-term contributions of this industry to their economies. Thus, the continuous enhancement of non-gaming attractions and facilities in these destinations (not just limited to the facilities within the casino/integrated resorts) will be a crucial part of their development agenda.

9.2.1 Arguments and evidence for changes

When advancing the development of the non-gaming sectors, experiences from the progress and success of Las Vegas are indeed a valuable reference for the casino tourism destinations in the ESEA. In particular, it is argued that "[g]aming is the economic engine of tourism in Nevada, but it is not the hook. It is the non-gaming attractions that are financed from gaming dollars that keep that state growing and the economy healthy" (Wynn 1993, 6). According to this argument, ensuring that the profit derived from casino gaming would be reasonably and appropriately reinvested to retain and elevate the attraction of a destination is the most important guiding principle. In practice, it is worthwhile to note that the development of non-gaming elements is unique to Las Vegas based on its geographical location, the natural resources and social and cultural structures in Nevada, as well as the attributes of the demand. This implies that to be strategic and effective, non-gaming facilities to be developed in the various ESEA destinations must have their own specific features.

Given the diversity in the contextual settings and tourism resources between the ESEA countries/regions, non-gaming elements that are found in these destinations will continue to have (and likely expand) their own features and compositions, thus differentiating their attractions from those of their rivals in the markets. In other words, since regional competition will increase, ongoing efforts for innovation to enhance the interplay between the gaming and non-gaming business segments will be essential for these destinations to retain their respective competitive edge in the long run. For example, although Macao largely relies on casino gaming, comprehensive non-gaming facilities invested by its IR operators through gaming income in the 2010s actually enhanced its real attraction and competitive edge in the region. According to a sampling analysis of casino tourism in Macao, Prentice et al. (2022, 465) find that compared to casino gaming, "non-gaming offerings, have a larger effect on customers' positive attitudes and behaviours." In addition, Prentice et al. (2022) further point out that as the mass visitors gain positive experiences, their motivation to revisit these properties is higher because they "view casinos as a destination with luxurious and unique architecture and facilities more in the sense of sightseeing" (ibid.).

Other than Macao, evidence derived from other major casino tourism destinations like Singapore (see, e.g., Henderson (2016)) and South Korea (see, e.g., Song et al. (2021)) also reveals that non-gaming facilities developed in the IRs contribute to their attractions, thus sustaining the flow of both local and foreign tourists who visit their properties and encouraging their overall spending. Nevertheless, as the consumption behavior of the masses for casino tourism trips has changed over time in different countries with a different tourist population (which is similar to the case of the Chinese millennials and Gen Z), the IR operators need to make continuous and innovative updates to the type of non-gaming facilities offered.

For example, according to another sampling analysis for the case of South Korea, Song et al. (2021, 7) confirm that the different forms of non-gaming "recreation facilities play a critical role in attracting more tourists to the casino-based resorts." Nevertheless, they also find that while traditional "skiing and golfing facilities operating by many large-scale resorts may not be an appealing element of IRs nowadays" (ibid., 8), visitors are increasingly welcoming family-oriented facilities like theme and play parks. On the other hand, when spending on trips to IRs, local Koreans "showed strong preferences for medium-sized casinos" (ibid.) rather than large-scale gaming venues. Indeed, this attitude may reflect their social concerns about over-gambling and the related problems (as explicated in Chapter 6). To this, they also express their desire that the Korean government allow more flexibility for local Koreans to enter the IRs in different tourism destinations instead of only the remotely located Kangwon Land under existing laws.

9.2.2 Non-gaming elements in a broader context

The role of the non-gaming elements established within IRs and their interplays with local tourism resources/attractions (including natural, historical and cultural ones) outside the IRs are depicted in Figure 9.1 with reference to the aforementioned argument on the importance of the success of casino tourism development in Nevada (mainly Las Vegas) and similar initiatives that have been adopted by some of the major destinations in the ESEA in the 2010s. Conventionally, most of the attention on the progress of casino tourism in the various ESEA destinations has been on the interactions between the gaming and non-gaming facilities of the IRs (i.e., *a* in Figure 9.1). To ensure that the IRs contribute to the overall and long-term development of a destination and its competitiveness, more efforts need to be made to boost cooperation between the IRs through their non-gaming facilities and other local tourism sectors (i.e., *b* in Figure 9.1). This will strengthen the overall attraction of the destination and generate a win-win situation. Actually, positive feedback effects (as shown by *c* in Figure 9.1) are produced to sustain the attraction of the destination as a whole, thus ensuring the inflow of tourists for the IRs and their growth.

As explicated in the earlier chapters of this book, the continuous progress of casino tourism in the more developed destinations like Macao (Chapter 4), Singapore (Chapter 5), South Korea and Malaysia (Chapter 6) and the Philippines (Chapter 7) in the 2010s and onward as well as the potential development in Japan (Chapter 8) well reflect that they are revamping the role and composition of the

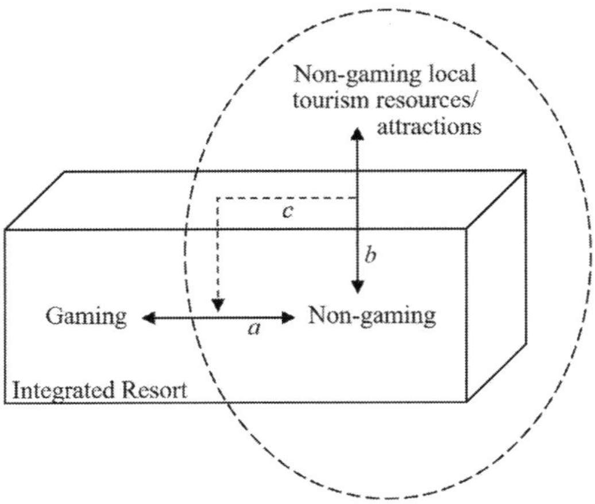

Figure 9.1 Role of non-gaming elements in a casino tourism destination

Source: Author

non-gaming elements not just for the development of the IRs but the broader context for the overall development of their destinations. Under these dynamics, the less-developed and potential destinations in the SEA (as described in Chapter 7) will also likely move in the same direction, although their direction might be influenced by their existing social and political settings. Otherwise, the real development of these destinations will be inhibited and their business performance greatly influenced by the further advancements and changes in government policies in the neighboring countries. Actually, these changing forces will most likely have unfavorable impacts on the less competitive destinations which remain loosely regulated and only offer some basic and monotonic non-gaming facilities like hotels, restaurants, retail shops, small-scale shows etc.

9.3 Dynamics and uncertainties ahead

Although casino tourism in the ESEA was hit hard by the COVID-19 pandemic, the development of the industry still retained momentum. Actually, it is observed that while the industry practitioners adopted a positive attitude and took action to cope with the unprecedented shocks to their businesses, the governments in the different destinations indeed continued to proceed with their agenda and make plans for the development of the casino tourism industry in their own country. Thus, the recovery of this industry in the various destinations is on the right track as anticipated. Also, new projects and changes have started to unfold in some of the destinations following the end of the pandemic. Thus, the development of casino tourism will likely continue despite that its business organization and routines have to go through some crucial changes in the near future owing to the forces discussed in the two previous sections.

To a large extent, the dynamics generated from the existing IRs and their potential expansion (including new properties in the existing destinations and newly developed destinations as elaborated in the previous chapters) will ensure the significance of casino tourism in the ESEA for a while and increase their attraction. Besides, the lenience of some of the governments in allowing their local citizens to spend time in the casinos in their own country may create new impetus to retain the vitality of those casino tourism destinations even with the decline of the traditional high-rolling gaming business segment. However, it is an unarguable fact that the development of casino tourism in the ESEA will face a crucial stage of transformation in the 2020s and even the 2030s. Parallel to the influences from the changes in the "China factor," changes in political inclinations and hence government policies in the various individual destinations will likely increase the uncertainty of the progress of this industry. In principle, the less developed destinations in the SEA will face greater uncertainties than the more developed destinations.

9.3.1 The dynamics

As a result of the dramatic development of casino tourism in the various countries/regions in the ESEA through the establishment of large-scale IRs in the 2000s and

2010s, iconic properties with comprehensive gaming and non-gaming facilities/entertainment and high-quality services will allow them to retain their attraction. At the same time, ongoing projects to renovate and expand existing IR properties will unarguably be helpful to sustain their development. For example, as indicated in the previous chapters, two IRs in Singapore – Marina Bay Sands and Resorts World Sentosa – will expand their existing properties and non-gaming facilities in the 2020s as part of their commitments in the renewed licenses of 2019. In South Korea, a new IR called INSPIRE Entertainment Resort opened at the end of 2023 in Incheon, which is a well-known and easily accessible tourism destination close to Seoul as opposed to Kangwon Land, which is located in a more remote region. Besides, discussions/debates have also ensued in Malaysia about whether an additional casino license could be approved for promoting tourism development in a tourism region called Sarawak.

In the world's casino capital, Macao, ongoing efforts will indeed be carried out by the IR operators to boost the interplay between the gaming and non-gaming segments even though its existing facilities for casino gaming are in principle restricted under the new gaming law in 2022 (however, reasonable adjustments in the number of gaming devices are possible which are subject to government review and approval). In addition, their commitments under the new gaming licenses to further the development of tourism in the city will undoubtedly generate new dynamics for the progress of casino tourism as a whole in Macao. On the other hand, the process of the privatization of PAGCOR and new projects to be carried out in the Philippines will also add to the new features of the markets. Accordingly, it is foreseeable that competition and potential competition within a casino tourism destination and between some of the neighboring destinations will increase. Nevertheless, competition between large-scale IRs may actually enhance the overall attraction of this industry in the ESEA. This is because making continuous innovations is a common strategy of IR operators to retain their competitive edge in the markets.

To the less-developed casino tourism destinations in the SEA (i.e., Vietnam, Cambodia, Laos and Myanmar), changes in the external environment/forces will indeed compel them to revamp their industrial organizations and improve their existing regulatory systems so that long-term development will be more promising. Actually, following the amendment of Article 303 of the Criminal Law of the People's Republic of China in 2021 and the actions taken accordingly, the governments of the various ESEA casino tourism destinations have indeed started to adjust and/or introduce new policies that improve the supervision and integrity of their gaming business. In this process, the governments of the less-developed destinations will likely (and have to) reconsider the structure and organization for the development of their casino tourism, which would truly contribute to the long-term benefits for their economies and societies. Besides, the strategy taken by the Thai government to legalize its casino gaming may not simply be exerting pressure on its neighboring countries which have long been targeting the Thai patrons but also generate positive dynamics so that existing casino tourism operators in those countries face potential competition, thus ameliorating the controversial segments of their businesses.

Last but not least, suppose that the efficiency of the regulatory systems in the ESEA casino tourism destinations will increase on an ongoing basis even if their progress may not be the same, with some of them slower than others. This trend implies that the markets will be better regulated over time. Therefore, despite the fact that the monetary benefits long associated with irregular and controversial casino gaming activities will decline, turnovers from the better-regulated business segments will indeed rise since there will be potentially more mass casino tourism visitors with higher income. Consequently, this trend will generate the essential force to ensure a more stable and sustained growth for the industry.

9.3.2 The uncertainties

In practice, there are still uncertainties even though the vitality of casino tourism and its transformation in the ESEA in the 2020s and 2030s are basically ensured by the aforementioned dynamics. To a large extent, these uncertainties mainly come from the possible changes in government policies at different levels. First, as the increasing number of Chinese outbound tourists and their spending have become a major force that has accelerated the development of casino tourism in the various ESEA destinations in the 2010s, any revised policies of the Chinese government that restrict or discourage its local citizens from spending on gaming activities when they visit casino tourism destinations will undoubtedly have unfavorable impacts on the business performance of the industry. Nevertheless, the degree of the impact will largely depend on the actions actually taken by the Chinese government under different circumstances and in different periods of time.

As argued in the first section, the amendment of the Criminal Law of the People's Republic of China in 2021 and actions taken immediately thereafter have conveyed clear messages to the industries and governments in the various destinations that the irregular/illegal business activities which have long been stifling the genuine development of casino tourism have to be resolved. Nevertheless, other possible ongoing actions and details of the actions that the Chinese government may have carried out can hardly be anticipated, which would also depend on the reactions of the industry practitioners and governments in those destinations. In other words, the tug-of-war between the Chinese government and the actual changes made in the casino tourism destinations will create ongoing forces that inhibit the prospects of this industry in the ESEA.

On the other hand, owing to the foreseeable economic pressure on the Chinese economy in the second half of the 2020s (or even the beginning of the 2030s) and geopolitical tensions between China and some of the ESEA countries, the likelihood that the Chinese government may introduce policies under consideration to further restrict and monitor the spending of Chinese outbound tourists on gaming cannot be neglected. If this occurs, the anticipated progress of casino tourism, including the development of the mass market segment in various destinations, will inevitably be affected. In practice, the uncertainties cannot be eliminated, although a scenario analysis can be conducted to gauge the possible effects before the policy changes.

Other than the changes in the policies of the Chinese government which will influence a wide range of the markets in the ESEA, changes in the policies by the governments of existing and potential casino tourism destinations will also add to the uncertainties in the development of this industry. For example, the desire of the Thai government to push the legalization of its casino gaming in the mid-2020s forward has not simply excited the investors (IR operators) but indeed threatened the sustained growth of casino tourism of its neighboring destinations to which the Thais have long been their patrons. However, the ultimate impacts may still be subjected to the actual layout of the industry which will be in the hands of the Thai government. Most importantly, how would the business activities be regulated in practice? Besides, the reactions of governments and industry practitioners in the neighboring destinations may further complicate the development of this industry in the Indochina region of the SEA.

In a nutshell, given the attributes of demand for casino tourism in various communities in the ESEA and properties of supply of the related services, the prospects of this industry in any one destination will be largely determined by the interplays between its government policies and the industry practitioners, as well as the policy impacts from the governments in neighboring countries under the consideration of their own political interests and benefits. Therefore, when planning the future of this industry, industry practitioners need to be proactive in assessing the possible changes in public interests (government inclinations) in the various destinations and proceed with caution.

9.4 Conclusion and remarks

The comprehensive evidence in the ESEA shows that even though casino tourism is in principle an industrial development strategy to boost tourism spending by providing casino gaming services as an alternative activity for tourists, the frontiers being developed for the practices of this industry are quite diverse between various destinations. Basically, advocates for the development of casino tourism mainly argue for its economic benefits. On top of the gaming and non-gaming revenues that can be derived from tourists within the casino hotels/resorts or integrated resorts, investment and employment opportunities and other positive spillover effects that can be generated by this industry for the local economy are commonly emphasized.

However, it is an indisputable fact that casino gaming is usually associated with various socially and politically undesirable activities or even creates the environment to breed and accelerate adverse social impacts both internally and externally. In practice, the degree of these negative spillover effects is inversely correlated, with the level of social development and positively correlated with the level of public corruption in a destination. Thus, to the development of any one casino tourism destination, its business performance and changes in different periods of time and its real contributions to the local economy are greatly influenced by the interplays between a set of social, cultural, political and economic (SCPE) forces that are unique to its society and their interactions with other external factors over time.

As explicated in the previous chapters, the attributes and compositions of the SCPE forces are quite diverse between the various ESEA countries/regions. Thus, the actual influences of each force and its aggregate impacts on the frontiers of casino tourism in these destinations are quite different. Given the internal forces, the specific external dynamics which actually led to the excitement and rapid expansion of casino tourism in the ESEA in the 2000s and 2010s are unarguably resultant of the special "China factor." In this period, however, it is also identified that the practices of casino tourism in the loosely regulatory environments of some of the destinations actually boosted the outflow of capital from mainland China and accelerated corruption and various adverse social impacts in the country. Thus, changes in the inclination of the Chinese government from the mid-2010s and revision of its criminal law in 2021 to restrict local citizens in mainland China from spending on casino tourism trips have clearly generated another essential set of dynamics which have altered the existing frontiers and path of development of most casino tourism destinations in the ESEA.

Together with the responses and changes introduced by other governments, evidence shows that progress of casino tourism in the ESEA has been on the path of transformation from focusing on high-rollers (heavy gamblers) to mass gaming tourists. From this, the interplays between the gaming and non-gaming business segments will likely be further strengthened in the various destinations by reinforcing the dynamics from the practices of the IRs, as well as their interactions with the local tourism sectors. Taking into consideration the iconic IR properties that have been (and will be) established in various destinations in the ESEA, their respective attractions and hence competitive edge will likely sustain their growth. On the other hand, owing to the different local tourism resources (including natural, historical and cultural resources) of the different countries/regions in the ESEA, regional competition in casino tourism may also be a kind of regional collaboration in some of the sub-regions in the long run.

As a final remark, it is important to emphasize that to ensure the sustainability of the development of casino tourism, a simple and important principle, but complicated and challenging practice for all of the destinations, is definitely the need for the governments to have an effective role at the outset. This role is largely reflected in their proactiveness and determination to formulate an effective regulatory system and make ongoing efforts to maintain and improve the efficiency of this system. To this, clear and appropriate incentives have to be conveyed from the practices of the regulatory systems (which comprise related laws and regulations and most importantly, their enforcement) to the markets. In other words, while market participants are being regulated, they should be provided with positive incentives to advance the development of this industry within a set of clearly defined rules (frontiers). When adjustments and improvements are needed in congruence with the changes (or potential changes) in the social, political and economic environments, effective communication between the different stakeholders in this industry is necessary to ensure its development and positive contributions to the economy.

Notes

1 https://www.statista.com/statistics/236765/total-population-of-china/
2 The general consensus is that the "[m]illennials were born between 1981 and 1996 while members of the Gen Z were born between 1997 and 2012" (Gorynski 2024).
3 Chinese citizens in mainland China need to obtain a government-approved visa to travel outside the country.
4 *China Statistical Yearbook* (National Bureau of Statistics of China 2002, 649; 2023, 575).
5 See, for example, https://www.theglobaleconomy.com/rankings/tourist_departures/. Due to the travel restriction policies imposed by the Chinese government from 2000 to 2022 to cope with the COVID-19 pandemic, the number of Chinese outbound tourists slumped during this period of time.
6 An unofficial English translation of Article 303 is available under section "36. Amend Article 303 of the Criminal Law to read" in China Law Translate 2020.

References

AGBrief Editorial. 2022. "Macau Urges Tourism Diversification as China Crackdown Intensifies." *Asia Gaming Brief* (July 25, 2022). Available at https://agbrief.com/news/macau/25/07/2022/macau-urges-tourism-diversification-as-china-crackdown-intensifies/. Accessed August 14, 2022.

Bao, Jigang, Xin Jin and David Weaver. 2019. "Profiling the Elite Middle-Age Chinese Outbound Travellers: A 3rd Wave?" *Current Issues in Tourism* 22 (5): 561–574.

China Law Translate. 2020. "P.R.C. Criminal Law Amendment 11." (December 26, 2020). Available at https://www.chinalawtranslate.com/en/criminal-law-amendment-11/. Accessed December 14, 2021.

Gorynski, Max. 2024. "Millennials vs. Gen Z: Differences in Customer Service Expectations." *Zendesk Blog* (February 22, 2024). Available at https://www.zendesk.com/blog/millennials-vs-gen-z-customer-service-expectations-compare/. Accessed February 28, 2024.

Henderson, Joan C. 2016. "Integrated Resorts and Tourism: A Singapore Perspective. *Asia-Pacific Journal of Innovation in Hospitality and Tourism* 5 (2): 191–208.

Huang, Linqian. 2023. "Analysis of the Changing Behaviours of Chinese GenZ Towards Casino Tourism: Implications to the Asian Markets." *Manuscript* (presented in the 18th International Conference on Gambling and Risk Taking, May 24, 2023). Abstract available at https://digitalscholarship.unlv.edu/gaming_institute/2023/May24/27/. Accessed June 24, 2023.

Meneses, João Paulo. 2022. "Special Report: Chinese Millennials." *Macau Business* (October 2022): 48–76.

Moura, Nelson. 2021a. "China's Supreme People's Procuratorate Warns of High Incidence of Illegal Gaming Operations with 'Out-of-Borders' Elements." *Macau News Agency* (November 29, 2021). Available at https://www.macaubusiness.com/chinas-supreme-peoples-procuratorate-warns-of-high-incidence-of-illegal-gaming-operations-with-out-of-borders-elements/. Accessed December 14, 2021.

Moura, Nelson. 2021b. "Gaming Operators, Junkets, and Private Citizens Should Carefully Study New Mainland Penal Code Amendments – Report." *Macau News Agency* (November 30, 2021). Available at https://www.macaubusiness.com/gaming-operators-junkets-and-private-citizens-should-carefully-study-new-mainland-penal-code-amendments-report/. Accessed December 14, 2021. (Ch04)

National Bureau of Statistics of China. 2002 and 2023. *China Statistical Yearbook* (in Chinese and English). Beijing, China: China Statistic Press.

Prentice, Catherine, Chunli Ji and Kefang Li. 2022. "The Influence of Gaming and Non-gaming Experience on Customers' Response to the Casino." *Journal of Vacation Marketing* 28 (4): 455–468.
Purbrick, Martin. 2024. "The China Crackdown on Cross Border Crime and Underground Banking." *The Asian Crime Century* (January 9, 2024). Available at https://asiacrimecentury.substack.com/p/the-china-crackdown-on-cross-border. Accessed February 4, 2024.
Song, HakJun, JunHui Wang and Seong Ok Lyu. 2021. "Willingness to Pay for Casino-based Integrated Resorts: A Choice Experiment." *Journal of Destination Marketing &* Management 19: 100555. https://doi.org/10.1016/j.jdmm.2021.100555
The Economist. 2022. "China's Cash Gift to the Rest of Asia." (December 10, 2022): 51.
Wilhelm, Kelsey. 2024. "Under the Scope: Cambodia Tightens Its Regulatory Belt, Aims to Appease China." *Asia Gaming Brief* (March 6, 2024). Available at https://agbrief.com/intel/deep-dive/06/03/2024/under-the-scope-cambodia-tightens-its-regulatory-belt-aims-to-appease-china/. Accessed March 7, 2024.
Wynn, Steve. 1993. "Legalized Gaming: Winners and Losers." *WLC Annual Meeting* (December 1993): 6.

Additional readings

Global Expat. 2023. "Bullish Outlook for Gaming-led IR Development in Asia." (August 22, 2023). Available at https://globalexpatrecruiting.com/bullish-outlook-for-gaming-led-ir-development-in-asia/. Accessed August 31, 2023.
Ho, Hong Wai. 2022. "Casino Development and Regulation in Asia: The Experience of Macau and Singapore for Japan and Other Destinations." *Asian Education and Development Studies* 11 (3): 505–520.
McCartney, Glenn. 2020a. "Securing Chinese Mass Market Visitation to Cotai's Integrated Resorts (IRs): Determinants of Gaming and Non-gaming Attributes That Influence IR Selection." *Tourism and Hospitality Research* 20 (3): 317–330.
McCartney, Glenn. 2020b. "Integrated Resort and Casino Tourism: A Global Hospitality Trend But a Sure Win?" In *The Routledge Companion to International Hospitality Management*. Edited by Marco A. Gardini, Michael C. Ottenbacher and Markus Schuckert, 199–210. New York, NY: Routledge.

Index

Made in the USA
Las Vegas, NV
19 September 2026

686df98c-befb-41ee-9364-adbdb4409d8eR01